Simple Kitchen Skills and Recipes

Mari Smith

Edward Arnold

© Mari Smith 1982
First published 1982 by
Edward Arnold (Publishers) Ltd
41 Bedford Square, London WC1B 3DQ

Reprinted 1983

British Library Cataloguing in Publication Data
Smith, Mari
  Simple Kitchen skills and recipes.
  1. Cookery
  I. Title
  641.5   TX717

ISBN 0-7131-0566-6

All Rights Reserved. No part of this publication
may be reproduced, stored in a retrieval system,
or transmitted in any form or by any means,
electronic, mechanical, photocopying,
recording or otherwise, without the prior
permission of Edward Arnold (Publishers) Ltd.

Printed and bound in Great Britain
at The Pitman Press, Bath

Contents

1 Equipment in the kitchen	4
Safety. Hygiene	4
Knives	6
Ham and cream cheese open sandwich	7
Preparing fruit	8
Fresh fruit salad	9
Preparing vegetables — chopping and dicing	10
Vegetable soup	11
Measuring solids	12
Chocolate crispies	13
Measuring liquids	14
Pear mouse	15
The liquidiser	16
Milk shakes and drinks	17
The refrigerator	18
Peach melba	19
2 How food is cooked	20
The cooker	20
Boiling	22
Stuffed eggs	23
Grilling	24
Welsh rarebit	25
Poaching	26
Egg Florentine	27
Frying	28
Fried sandwiches	29
Cooking in the oven	30
Flapjack	31
Stewing	32
Quick sausage hot-pot	33
3 The food we eat	34
Flavour and texture	35
Meat	36
Meat loaf	37
Fish	38
Stuffed baked fish	39
Eggs	40
Scrambled egg	41
Pulse vegetables	42
Special baked beans	43
Cereals	44
Muesli	45
Flour	46
Soda bread	47
Dietary fibre	48
Wholemeal scones	49
Starch	50
Fruit fool	51
Cheese. Sauces	52
Macaroni cheese	53
Fruit and vegetables	54
Stuffed tomatoes. Coleslaw	55
Grow your own salad	56
Mixed salad. Dressing	57
Fruit drinks	58
Fresh lemon or orange drink. Citron pressé	59
Timing and table laying	60
Preparing a meal — breakfast	61
Convenience foods	62
Hamburgers	63
Magic Word Square	64

1 Equipment in the kitchen

The kitchen is the room where food is prepared, cooked, and sometimes eaten. Most equipment for these activities is kept in the kitchen, from small items such as knives, to large pieces of electrical equipment such as cookers and refrigerators. The kitchen is often a warm, cosy room where the family like to gather, but it can also be a dangerous place. Here are some points to help to make it a safe place.

Kitchen safety

Electrical equipment should be checked to make sure that plugs are fixed firmly and leads do not become frayed. Do not touch plugs and switches with wet hands, keep plugs away from water.

If there is a smell of gas, find out where it is coming from, open the window until the smell clears. If you cannot find the source of the smell, ring the gas board at once.

Never leave a pan of oil or fat unattended on the heat, it can easily catch fire.

Keep the floor clear of toys, bags and rugs, that could be tripped over.

Wipe up spills that could make the floor dangerously slippery.

Keep dangerous chemicals, e.g. bleach, away from children.

Learn to use equipment correctly so that cuts and burns do not happen. But keep a first aid box in the kitchen containing waterproof plasters, antiseptic cream, cotton wool.

First aid for cuts: wash in cold water, dry, cover with a plaster. Cuts must always be kept covered when handling food.

First aid for burns: put the burn in cold water at once, dry and leave uncovered.

The kitchen below would be a very dangerous place! How many careless, dangerous items can you find?

Kitchen hygiene

No one would like to eat food that has been prepared in dirty surroundings; it could spread food poisoning and other diseases. Therefore it is important that both you the cook, and the kitchen are clean. Cleanliness in the kitchen is known as kitchen **hygiene**.

You in the kitchen
a) Before you begin.
   Wash your hands.
   Wear a clean apron.
   Tie back long hair.
   Use a clean dish cloth and tea towel.
   Check that work tops are clean.

b) As you work.
   Wrap up rubbish and peelings, throw away.
   Do not block the sink with peelings.
   Leave dirty pots and pans to soak in soapy water. Protein foods, e.g. egg, milk, soak more easily in cold water. Other foods benefit from a hot soak.
   Wash up as you work.

c) When you finish.
   Wash over all surfaces.
   Wash up using hot water in a washing up bowl with a little detergent – not too much.
   Wash the cleanest things first, the dirtiest last.
   If possible, leave dishes to drain dry before putting away.
   If a tea-towel is used it must be clean or it will only spread dirt on to the newly washed items.
   Sweep the floor, then look at the kitchen and ask yourself whether you would be happy to work in it and to eat food prepared there.

New word
Hygiene – cleanliness and health

Knives

When preparing food, it is important to use the correct utensil for each task. This makes it quicker, safer and easier. Knives play an important part in food preparation, so should be used correctly.

Vegetable knife
This small sharp knife is used for chopping vegetables and small pieces of food. Use it on a chopping board to prevent scratching the table top. Dirt and bacteria become trapped in the scratches and may spread food poisoning so the board should always be washed after use.

Palette knife
This is a blunt knife with a rounded blade. The blade is flexible so it bends easily. This makes it useful for many tasks, e.g. spreading, scraping, lifting omelettes, eggs and pancakes, mixing and stirring.

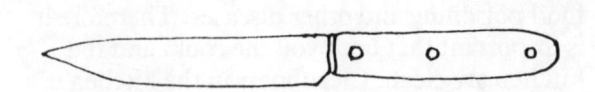

Cook's knife
This should be kept sharp for cutting meat and large pieces of tough food. Hold it correctly, cutting firmly down on to the board.

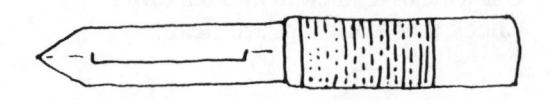

Vegetable peeler
Sometimes called a safety peeler, it is used to peel fruit and vegetables thinly, preventing waste and saving vitamins. The blade is protected, making it safer than a knife.

Things to do

1. Draw a diagram of a cook's knife, a vegetable knife and a vegetable peeler. Label each one and suggest two uses for each.
2. Copy the following sentences into your book and fill in the correct word from the list underneath:
   a) The blade of a knife is made from ——.
   b) The handle is made from —— or p——.
   c) The blade of a palette knife is b—— and f——.
   d) The blade of a vegetable knife is ——.
   plastic, sharp, steel, blunt, wood, flexible
3. Draw a simple garnish to be made from cucumber. A green herb often used to garnish food is ——.
4. Suggest two new recipes for open sandwiches using some of these ingredients:
   tomato, egg, lettuce, chives, onions, tuna fish, sardines, watercress, apples, walnuts, cream cheese.
5. There are many gadgets that will do the job of a knife, e.g. apple corer, auto chop, chipper, egg slicer, parsley chopper, garlic press. Compare each of these with the use of a knife.
6. Experiment to choose the better utensil. Take two potatoes of the same weight. Peel one potato with a peeler. Peel the other carefully with a knife.
   Weigh each potato after peeling. Complete the following:
   weight of potato – ——
   weight of potato after using knife – ——
   weight of potato after using peeler – ——
   The potato peeled with —— lost less peel, therefore there is less waste if this is used.

Ham and cream cheese open sandwich

Ingredients
for 4 people
4 slices of wholemeal bread
75 g cream cheese
12 g butter
1 tsp made mustard
4 slices cooked ham
8 slices cucumber

Utensils
bread knife, bread board, palette knife, basin, teaspoon, vegetable knife, chopping board, serving plate

Method
Collect ingredients and utensils.

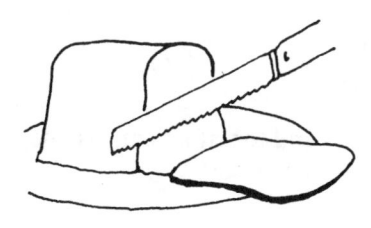

1 Cut 4 slices of bread using a bread knife.

2 Spread with butter. Remember to take butter from the refrigerator ½ hour before use to soften.

3 Beat the cream cheese and mustard together in a basin.

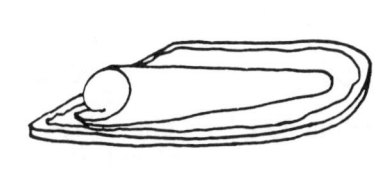

4 Spread cheese mixture evenly on each slice of bread. Roll up slices of ham into cornets, arrange on the bread.

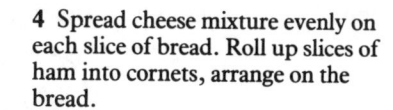

5 Use a vegetable knife on a chopping board to slice the cucumber into thin circles.

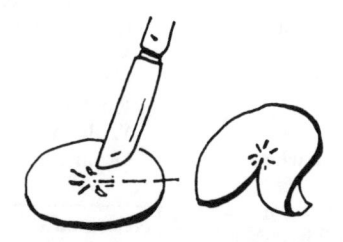

6 Make cucumber twists.

7 Garnish the sandwiches with the cucumber twists and serve as soon as possible with a glass of milk or fruit juice for lunch or supper.

Sandwiches are a quick, useful food to prepare. Although they seem simple, they can be made with a variety of sweet and savoury fillings and many types of bread and crispbread.

The open sandwich comes from Denmark and has bread only at the bottom; the filling lies on top and therefore it is important that it looks attractive.

The traditional English sandwich was 'invented' by the Earl of Sandwich. Look at a map of England to find the town of Sandwich.

New word
Garnish – to decorate food with something edible to make it look attractive

Preparing fruit

Oranges
The orange has a thick skin including a bitter white pith which must be removed completely.

Use a vegetable knife or one with a serrated blade. Hold the orange firmly on the board. Use a sawing action to remove all the skin and pith, exposing the flesh.

Loosen each segment by cutting down the skin separating them.

Apples/pears
Peel thinly using a vegetable peeler. Rub the cut surfaces with lemon juice to prevent them becoming brown.

Cut the fruit into halves, then quarters. Remove the core and pips. Slice thinly, cutting the slices into small pieces for fruit salad so that each piece will easily fit on a teaspoon.

Bananas
Remove peel, slice across and coat with lemon juice. Prepare bananas last to prevent them becoming soft and brown.

Other fruit
Wash and dry the fruit. Remove skins if necessary. Halve grapes or cherries and remove the pips. The skins of peaches or apricots may be removed by **blanching**. To blanch, place the fruit in a basin, cover with boiling water for ½ minute only. Drain off the water and the skin will slip off easily.

The skins of almonds and tomatoes can also be blanched to remove them easily.

Orange baskets
Use a sharp pointed knife to cut around the centre of the orange in a zigzag line.

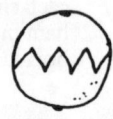

Cut through to the centre of the fruit, separate the two halves. Remove the pulp using scissors to loosen the core and scoop out the flesh with a teaspoon. Use the empty halves as containers for fruit salad.

Things to do

1. Draw a diagram showing how to
   a) prepare an orange for fruit salad
   b) prepare an apple for fruit salad
2. Rearrange the letters in these words to give the names of other fruits that can be used in fruit salad:
   a) acphe b) yperbsrra c) tricoap d) rswrtybare e) yrherc.
3. Complete these sentences:
   a) Apples, pears and bananas become –r––– when peeled.
   b) To prevent this they are coated with l––––– –u–c–.
   c) The syrup must be –––– before the fruit is added.
   d) The skin of peaches is removed by –––n–––––.
4. The following words are the names of varieties of fruit. Fill in the name of the correct fruit (the first one has been done for you).
   Cox's Orange Pippin — apple
   Victoria —
   Conference —
   Russet —
   William —
   Granny Smith —

New word
blanch – to cover with boiling water for 1–2 minutes

Fresh fruit salad

Ingredients
for 2–3 people
1 eating apple
1 pear
1 banana
50 g black or green grapes
1 orange
1 lemon
250 ml water ⎫ syrup
50 g sugar ⎭

about ½ kilo

Utensils
saucepan, jug, wooden spoon, vegetable knife, peeler, chopping board, lemon squeezer, basin, serving dish

Method
Collect ingredients and utensils.

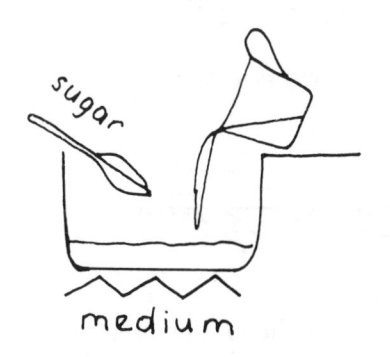

1 Place sugar and water in saucepan. Dissolve over medium heat. Bring to the boil and boil for 1 minute.

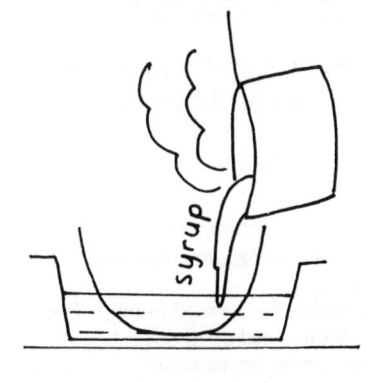

2 Leave to cool. It may be cooled quickly by being placed in a basin in a bowl of cold water.

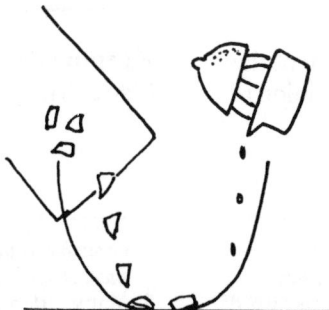

3 Prepare the fruit: peel the orange and divide into segments; halve the grapes and remove pips. Place in serving dish, reserving the juice from the orange.

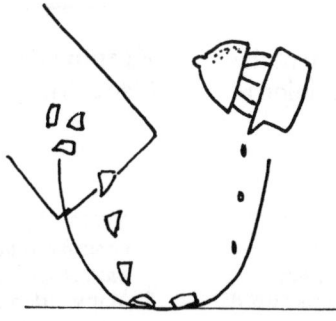

4 Squeeze the lemon juice into a basin. Peel and slice the apple and pear, drop into lemon juice so each piece is covered.

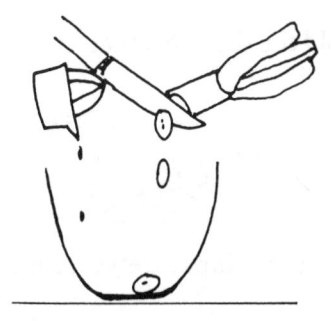

5 Peel and slice the banana, coat in lemon juice.

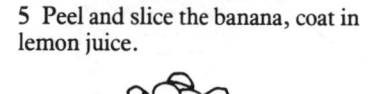

6 Place all fruit in the cold syrup. Chill in the refrigerator before serving. Serve in a glass bowl, individual dishes or a 'basket' made from the skin of an orange or grapefruit.

The syrup helps the fruit to keep its colour and preserves the vitamins.

Preparing vegetables — chopping and dicing

Using knives to chop and dice is a useful skill in food preparation. Always use a chopping board. Wrap up peelings before throwing them away. Dice are small even-sized cubes of food, e.g. in mixed vegetables, soups.

To dice vegetables: cut the food in half to give a flat surface, hold the flat surface down on the chopping board, cut into slices using a vegetable knife.

Hold the slices together, cut into strips or chips.

Hold the chips together, cut across into dice.

In this way, food can be diced quickly and evenly.

Chopping an onion
Use a vegetable knife to cut off the stem at the top. Leave the root end uncut. Cut the onion in half from stem to root, remove the skin, trim away the roots.

Place cut side of onion face down on board and slice leaving the root end uncut; this will help to hold the onion together.

Cut into strips, then across into dice.

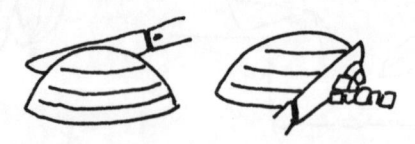

Chopping parsley
Use the larger cook's knife for this. Wash and dry the parsley. Remove leafy parts and throw away stems. Chop the leaves on a board holding the knife at the point and the handle.

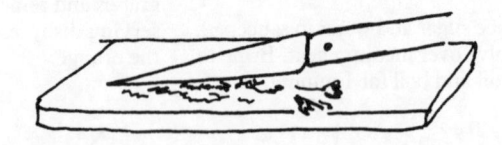

Keeping the point fixed to the board, move the blade firmly up and down on the parsley.

Things to do

1. Draw and label diagrams showing how to
   a) dice a potato b) chop an onion c) chop parsley
2. Complete these sentences, choosing the correct word from the list opposite:
   a) Dice are small, even —— of food.
   b) To dice vegetables, halve, then ——, chip, ——.
   c) Use a —— —— on a —— —— to chop and dice.
   d) A —— —— is used to chop parsley.
   e) Soup is covered and —— during cooking.

   vegetable knife, cook's knife, cubes, dice, simmered, slice, chopping board

3. Many different vegetables may be used in soup. The names of some have been mixed up in this list. Rearrange the letters to give the names of:
   Root vegetables that grow in the ground
   a) ekel b) rcarto c) nptiru d) nnoio e) spnarpi
   Vegetables that grow above ground
   a) mtaoot b) suoomrhm c) yclere d) sguaarpsa

Vegetable soup

Ingredients
for 3–4 people
1 carrot
1 onion
1–2 sticks of celery } about 800 g
1 leek
½ turnip
1 potato
1 tblsp cooking oil
600 ml stock (1 stock cube + 600 ml water)
1 tsp tomato purée
salt, pepper, parsley

Utensils
saucepan with lid, wooden spoon, tablespoon, chopping board, peeler, vegetable knife, colander, measuring jug, soup bowls

Method
Collect ingredients and utensils.

1 Wash the vegetables. Peel and chop the onion. Peel and dice the carrot, turnip and potato. Wrap up peelings and throw away.

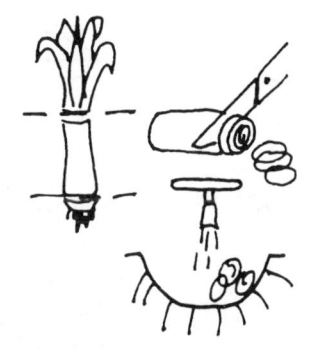

3 Remove the green leafy part of the leek. Slice the white part across thinly. Separate the layers of leek and rinse well in a colander to remove dirt.

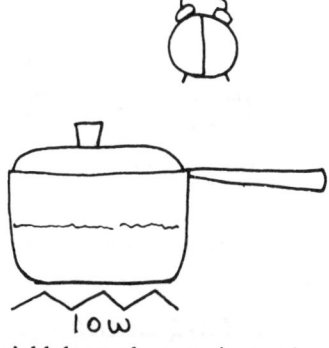

5 Add the stock, seasonings and tomato purée. Cover the pan and **simmer** for 30–40 minutes until the vegetables are soft.

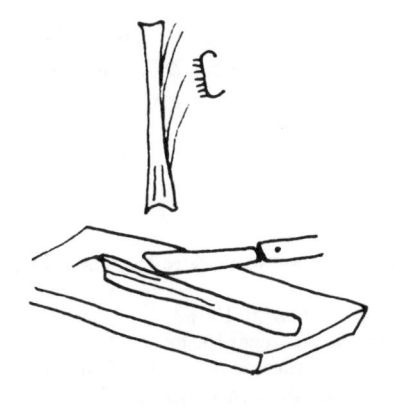

2 Scrub the celery and remove any tough, stringy parts. Cut down the centre and across into dice.

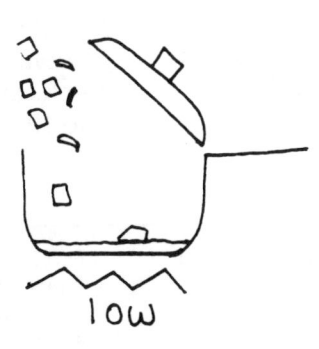

4 Heat the oil in a saucepan. Add the diced vegetables. Cover and fry gently for 5 minutes until they begin to soften.

6 Chop the parsley. Serve the soup in warm bowls sprinkled with parsley; grated cheese may also be served with it. Eat with toast, bread or rolls.

New word
Simmer – to cook gently below boiling point, covered with a lid

Measuring solids

The ingredients in a recipe must be carefully weighed and measured for good results. Solid foods are measured in grams and kilograms. These are written in shorter form: gram – g, kilogram – kg. These short forms are known as **abbreviations**.
1000 g = 1 kg   500 g = ½ kg

Measuring using scales
There are two main types of scales giving accurate measures.

balance scales

The scale pans balance evenly at the correct weight.

spring scales

wall scales

Check that the pointer is at 0 when the scale pan is empty, then add the ingredients until the needle reaches the weight needed.

Measuring using spoons
Special measuring spoons can be bought in various sizes, or a standard tablespoon may be used for dry foods. Three useful spoons are:

teaspoon tsp.
tablespoon tblsp.
dessertspoon dsp.

2 teaspoons = 1 dessertspoon
2 dessertspoons = 1 tablespoon

If a recipe needs 1 rounded tblsp, it means that there should be as much of the ingredient above the rim of the spoon as below it.

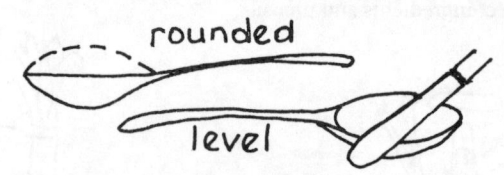

rounded
level

A level tblsp means the ingredients are levelled off with a knife, so they are flat with the rim of the spoon; ½ spoonful of any ingredient means half the level spoonful is used; ¼ spoonful means a quarter of the level spoonful is used. Some useful spoon measures:
1 rounded tblsp flour = 25 g
2 level tblsp sugar = 25 g
3 level tblsp icing sugar = 25 g
4 level tblsp grated cheese = 25 g
2 level tblsp fat = 25 g
1 level tblsp syrup = 25 g

Most hard fats are sold in blocks of 250 g, the block can be cut into 25 g pieces by dividing it into 10 even pieces, each one weighing 25 g.

Things to do

1. Draw a diagram of a tablespoon, a dessertspoon, and a teaspoon. Label them with their full names and suggest one use for each.
2. Give the abbreviations for these words:
   kilogram, gram, tablespoon, teaspoon, dessertspoon
3. Fill in the correct word to complete the following:
   a) 1 rounded —— flour = 25 g.
   b) 2 level tblsp sugar = ——.
   c) 2 tsp = ——.
   d) 2 dsp = ——.
   e) 4 tsp = ——.
   f) There are —— g in 1 kg.
4. Draw a diagram to show how to measure 25 g margarine.
5. Work out the following:
   a) How many tblsp flour weigh 100 g?
   b) How many tblsp sugar weigh 100 g?
   c) How many grams are there in 6 rounded tblsp flour?
   d) How many grams are there in ¼ kg?
   e) How many tblsp flour weigh ¼ kg?

New word
Abbreviation – the short form of a word

Chocolate crispies

Ingredients
(for 8 crispies)
25 g margarine
1 level tblsp drinking chocolate
1 tblsp golden syrup
6 tblsp cornflakes or rice crispies

Utensils
saucepan, plate, tablespoon, wooden spoon, teaspoon, knife, 8 paper cake cases, serving plate, d'oyley.

Method
Collect ingredients and utensils. Arrange cake cases on a plate.

1 Measure the golden syrup into a saucepan. To stop the syrup sticking, warm the spoon in hot water and use the hot wet spoon; the syrup will slide off easily.

3 Remove from heat. Stir in the cornflakes or rice crispies. Stir thoroughly to coat evenly with chocolate.

5 Leave in a cool place or the refrigerator until cold and set–at least ½ hour.

2 Place margarine and drinking chocolate in the saucepan. Dissolve over a low heat stirring with a wooden spoon. Do not boil.

4 Place a heaped tsp of the mixture in each case. Use a knife or spoon to help shape it. Try to avoid using fingers to poke into the mixture and resist the temptation to lick them.

Measuring liquids

Liquids are measured in litres and fractions of a litre called millilitres or decilitres. The following abbreviations are used:
litre – l
millilitre – ml
decilitre – dl

1000 ml = 1 l   10 dl = 1 l
500 ml = ½ l   5 dl = ½ l
250 ml = ¼ l   2.5 dl = ¼ l

Special measuring jugs can be bought. These have a scale on the side showing the liquid measure.

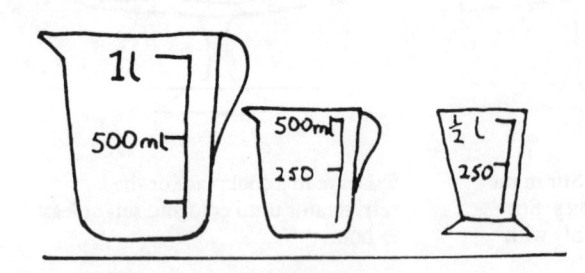

A milk bottle contains 1 pint (pt) liquid. This makes it a useful liquid measure if it is remembered that 1 pt = 600 ml or 6 dl, and ½ pt = 300 ml or 3 dl.

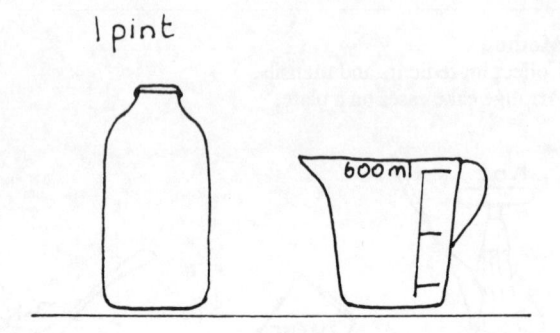

Another useful measure is the plastic spoon that comes with medicines; this holds 5 ml.

All pre-packed solid and liquid foods must be labelled showing the weight of the goods or the volume of liquid. The net weight of a product is marked on the label.

Things to do

1. Draw a 1 litre measuring jug. Mark the 500 millilitre and 250 millilitre levels.
2. Give the abbreviations for:
   litre, millilitre, decilitre, pint
3. Complete this table:

	dl	ml	
there are			in 1 litre
there are			in ½ litre
there are			in ¼ l
there are			in 1 pt

4. Choose the cheapest
   a) from the bottles of washing up liquid 
      5 l — £1·65   750 ml — 33p   1 l — 35p
   b) from the bottles of cooking oil 
      2 l — £1·28   1 l — 65p   900 ml — 75p   ½ l — 45p
   c) from the bottles of lemonade
      1 l — 55p   540 ml — 25p   740 ml — 45p

New word
net weight – the weight of the ingredients alone, without the weight of the tin or packaging

Pear mouse

Ingredients
1 packet lemon jelly
1 400 g tin pear halves
currants, angelica, cherries, blanched almonds
6–7 ice cubes

Utensils
measuring jug, wooden spoon, tin opener, kettle, 600 ml serving dish

Method
Collect all ingredients and utensils.

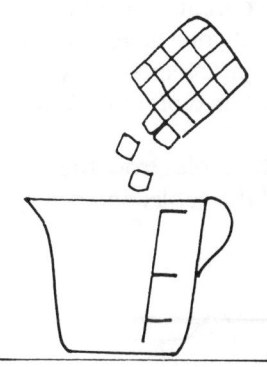

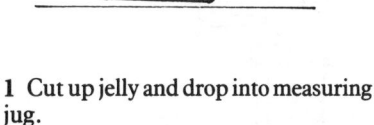

1 Cut up jelly and drop into measuring jug.

2 Cover with boiling water to come up to the 200 ml mark. Stir with a wooden spoon until completely dissolved.

3 Open the tin of pears. Strain the syrup into the jelly.

4 Cool the jelly quickly by adding 6–7 ice cubes, stirring to dissolve them thoroughly until the liquid reaches the 500 ml mark.

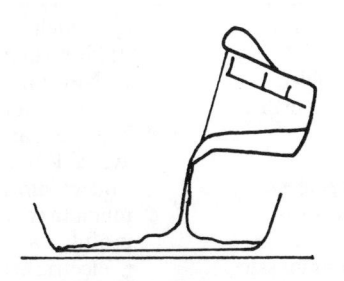

5 Pour into a serving dish. Leave to set in the refrigerator until firm.

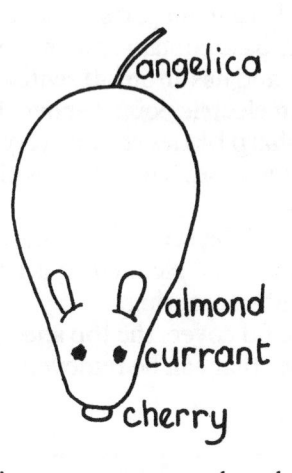

6 Arrange currant eyes, almond ears, angelica tail and cherry nose on each pear half.

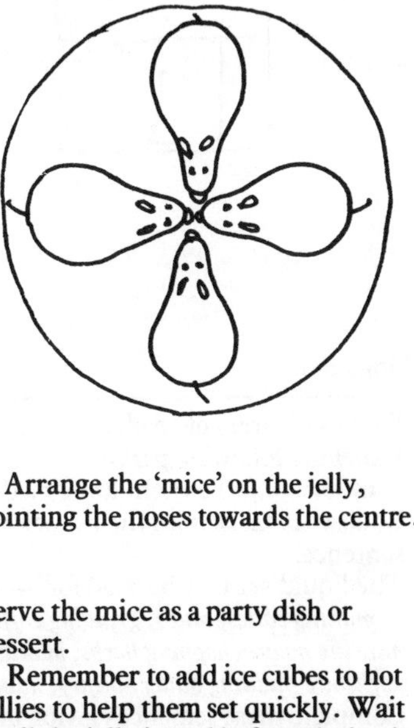

7 Arrange the 'mice' on the jelly, pointing the noses towards the centre.

Serve the mice as a party dish or dessert.
  Remember to add ice cubes to hot jellies to help them set quickly. Wait until the jelly is cool before placing in the refrigerator; never put hot foods in the refrigerator.

The liquidiser

The liquidiser is a useful piece of electrical equipment in the kitchen. It liquidises or blends food into a pulp and may be used to make drinks, soups and purées. It can also chop dry foods, e.g. nuts, cheese, breadcrumbs. It must be used with care.

There are many types of blender, but they all have three main parts: base, goblet, and lid.

The base contains the motor which is hidden inside and has an on/off switch which is plugged into an electric point. Above this three or four very sharp blades revolve very fast when the machine is on. These blades chop or liquidise the food.

The goblet screws on to the base and contains the food to be blended. It is usually marked in litres for measuring.

The lid covers the top and may have an inner stopper that can be removed separately. This allows small amounts of food or liquid to be added on to the revolving blades when the machine is on.

Using the liquidiser

a) Read the instructions with the liquidiser.
b) Place ingredients in the goblet, do not overfill. Cover with the lid and plug in. Switch on at the main switch, then at the on/off switch. Blend for the time needed, but do not run the machine for more than a minute at a time.
c) *Never* try to stir the contents when the liquidiser is on. *Never* put your fingers in the goblet; always use a scraper or spoon to stir when the machine is switched off at the main switch.
d) After use, remove the goblet from the motor, wash all parts of the goblet and dry thoroughly.

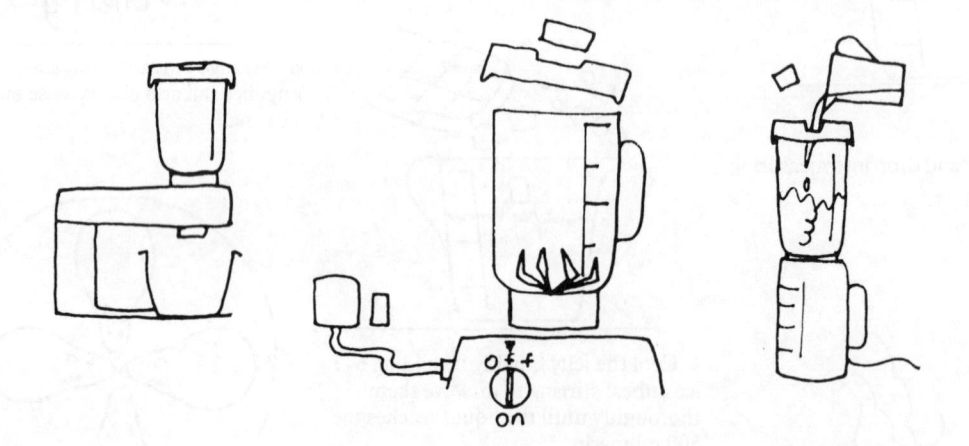

Things to do

1 Draw a diagram of an electric blender or liquidiser. Label the following parts:
   motor, on/off switch, knife blades, lid, goblet.
2 Choose six tasks from the list below to complete this sentence:
   The liquidiser may be used for ——
   making cakes, soups, grinding coffee, chopping nuts, mincing meat, chopping herbs, making breadcrumbs, yorkshire pudding batter, pastry, milk shake.
3 Complete these sentences giving instructions for safe use of the liquidiser:
   a) Do not touch plugs or switches with w—— h——.
   b) Do not o—— the goblet.
   c) Switch —— before stirring contents of goblet.
   d) Stir using a ——.
   e) Never put f—— in the goblet.
   f) Never put the m—— in water.
4 There are many other pieces of electrical equipment available to help with food preparation in the kitchen. Collect information about the following, find out how much they cost and suggest which would be most useful:
   electric beaters, potato peeler, coffee grinder, carving knife, tin opener, mincer, fruit juice extractor, chopper/slicer, food processor

Milk shakes and drinks

Banana milk shake

Ingredients
(for 2 people)
1 ripe banana
1 tsp caster sugar
300 ml milk
1 tblsp vanilla ice-cream

Utensils
blender, serving glasses, jug, tablespoon

Method
Collect ingredients and utensils.

Peel and chop the banana.
Place all ingredients in blender.
Blend on highest speed for 1 minute or until smooth.
Pour into glasses and serve.

Chocolate milk shake

Ingredients
300 ml milk
1 level tblsp drinking chocolate
1 tblsp ice-cream (optional)

Method
Place all ingredients in blender on high speed for 1 minute.

Many other flavours may be made in the same way, e.g. coffee (with 2 tsp coffee essence), blackcurrant (with 1 tblsp black currant syrup), strawberry (with 1 heaped tblsp strawberry ice-cream).

Egg flip

Ingredients
(for 1–2 people)
1 egg
300 ml milk
1 tsp honey

Method
Blend all ingredients for 1 minute. Serve chilled for a refreshing drink or quick breakfast.

Blender lemonade

Ingredients
(for 3–4 people)
1 lemon
1 litre water
2 tblsp sugar

Utensils
electric blender, vegetable knife, glasses, jug, tablespoon, sieve

Method
Collect ingredients and utensils.

1 Wash lemon and cut off pointed ends. Chop lemon into 6 pieces. Place in goblet with sugar. Blend on high speed for ½ minute. Switch off, add water, blend for 1 minute.

2 Switch off. Strain lemonade through a sieve into a serving jug.

The refrigerator

A refrigerator is useful for keeping food fresh as well as for chilling food before serving, setting jellies etc. The ice-box or evaporator is used to store frozen food, e.g. ice-cream. The coldest part of the refrigerator is the evaporator, usually found at the top. This is the only part of the refrigerator with a temperature below freezing point (0°C) where water can be frozen to make ice-cubes. Frozen food can be stored in the ice-box for varying lengths of time, depending on the temperature inside it. This is indicated by * markings on the door of the evaporator:

* * * coldest – frozen food will keep up to 3 months
* * less cold – frozen food will keep up to 1 month
* least cold – frozen food will keep up to 1 week

The evaporator is not as cold as a deep freeze; it is not used to freeze fresh foods, but to store frozen items.

Some foods can be cooked while still frozen, others must be thawed or defrosted before use. Frozen chicken must always be thawed before cooking. Never re-freeze foods that have been thawed; store in the refrigerator and use as soon as possible. The shelves inside the refrigerator are not as cold as the evaporator, but cold enough to keep foods fresh. The coldest shelf is at the top. The least cold part is inside the door and is suitable for eggs, milk and fats. Some foods will keep longer than others:

Food	Storage time	Where to store
raw meat	up to 3 days	top shelf
cooked meat	2–3 days	centre
raw mince	1 day	top shelf
raw fish	1 day	top shelf
milk	up to 3 days	door
cheese	up to 2 weeks	bottom shelf
salad vegetables	up to 4 days	bottom

Take care when using the refrigerator to store food correctly. Cover food to prevent it drying out, and to prevent smells spreading.

Never put hot food in the refrigerator. Opened tins of food should be emptied into dishes, not stored in the tin. When the refrigerator is defrosted, emptied and switched off, e.g. before going away, leave the door open.

Things to do

1 Draw the diagram of a refrigerator. Label it to show where you would store these foods:
butter, cheese, lettuce, fresh meat, frozen fish, milk, tomatoes, ice-cream

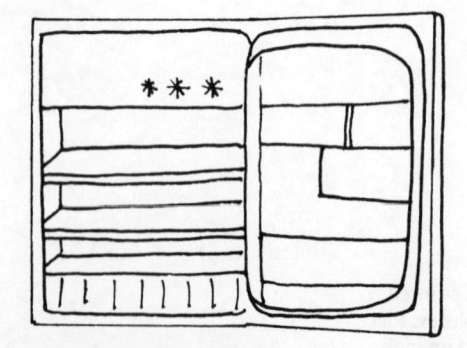

2 Complete the sentences, choosing the correct word from the words in brackets:

a) Never put —— food in the refrigerator. (*cold, hot, iced*)
b) Always —— food in the refrigerator. (*cover, heat, wrap*)
c) —— must be thawed before cooking. (*meat, fish, chicken*)
d) Do not refreeze food that has —— (*cooked, thawed, defrosted*)

3 Complete the chart to give the correct storage times for these foods in the refrigerator with a * * * ice-box.

Food	Length of storage time
ice-cream	
raw meat	
fish fingers	
mince	
lettuce	

Peach melba

Ingredients
(*for 2 people*)
2 ripe fresh peaches (or small tin peach halves)
100 g fresh raspberries
2 tblsp icing sugar
2 small blocks vanilla ice-cream
chopped nuts

Utensils
sieve, basin, vegetable knife, wooden spoon, tablespoon, dessert dishes

Method
Collect ingredients and utensils.

1 Blanch the peaches in boiling water to remove the skins, if wished.

2 Drain off the water, slit the skins and remove. Cut the peaches in half, twist the 2 halves in opposite directions to separate. Remove the stone.

3 Wash the raspberries. Make into a purée, adding the icing sugar to sweeten.

4 Place 2 tblsp ice-cream in each serving dish. Sandwich with a peach half on either side. Pour over the raspberry or melba sauce. Serve with a crisp wafer.

Making a purée
Food may be made into a purée in two ways:

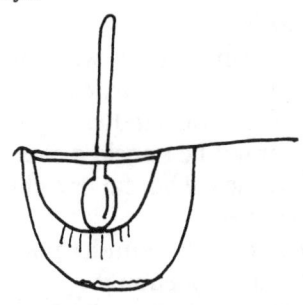

1 Place the soft food, e.g. fruit or cooked vegetables, in a sieve over a basin. Press the food firmly through the sieve using a wooden spoon. Scrape the purée from the base of the sieve using a clean tablespoon, throw away pips, skin or waste that will not pass through the sieve.

2 Use an electric blender. Place food in the goblet, blend on high speed for 1 minute until smooth.

New word
purée – a smooth pulp of food that has been sieved or liquidised, e.g. tomato purée

2 How food is cooked

The cooker

The fuels used to heat modern cookers are – natural gas, electricity, solid fuels, and calor gas. For centuries food was cooked on a spit over an open fire; now the most common fuels are gas and electricity. The two fuels may be combined in a split level cooker, e.g. gas hob with electric oven.

There are many types of cooker, but some basic safety rules apply to most of them.

The hob
Choose pans that are well balanced and fit well over the rings or burners to prevent waste of heat. Turn the handles of pans inwards so that they cannot be knocked over or become hot.

Never leave hot fat or a chip pan unattended on a cooker.

Lower the heat under pots and pans as soon as they are hot enough.

The grill
Adjust the grid in the pan to the height needed: near the heat for toast and thin foods, further away for thicker foods or those with a lot of fat, e.g. sausages. Watch the food under the grill – it cooks quickly and may burn easily. Draw out the pan to turn the food, do not poke around under a lighted grill.

If using an eye level gas grill, check that nothing is left on top that could catch fire, melt or become too hot.

The oven
Check that oven shelves are in the position needed before the oven is switched on and they become hot. Select the temperature needed, your recipe will tell you this. Light the oven 15–20 minutes before it is needed to allow it to heat to the temperature chosen.

The oven temperature is kept at the chosen heat by a device called a **thermostat**, a mechanism for controlling temperatures also found in refrigerators, irons, central heating systems and water heaters.

Always use oven gloves to protect your hands. Stand to one side of the oven door as it is opened, this will protect you from the rush of hot air rising from the open oven.

Switch off after use, wipe up spills and dirt while still warm.

The electric hob
Check that the main switch is on (do not touch the switch with wet hands). Never put your hands on the rings to check if they are on.

The electric oven
The heat is measured in degrees Farenheit (°F) or degrees Celsius (°C) in newer ovens. When the oven is switched on, a light appears near the oven switch showing that it is heating up. When the oven reaches the temperature you have chosen, the light goes out. The top shelf will be a little hotter than the others because hot air rises. However, if you are using a fan assisted oven, hot air is blown evenly around all the shelves.

The gas hob
Turn the gas knob on full and light the gas at once. Lower the flame to the height needed. Avoid draughts that may blow out the flame.

The gas oven
The heat in a gas oven is measured in numbers ¼–9. These are sometimes called gas marks and begin with the coolest at ¼, rising to the hottest at 9. This oven has three zones or areas of heat.

New word
thermostat – a mechanism for controlling temperature, e.g. in an oven

The top shelf is hottest, the middle shelf is the heat set by the oven knob, and the lowest shelf is the coolest. This is very useful as it means that three different dishes needing different heats can be cooked at the same time.

To light the gas oven:
a) Open the door.
b) Turn the gas on full.
c) Light the gas at once.
d) Close the door when the gas is alight.
e) Turn oven knob to gas mark needed.

Oven chart

Gas no. (mark)		Electricity °F	°C
2	cool — meringues	300	150
3	warm – milk puddings, custards	325	170
4	moderate – stews, casseroles	350	180
5	fairly hot – cakes, biscuits, baking	375	190
6	moderately hot – roasting, baked potato, pastry	400	200
7	hot – Yorkshire pudding, flaky pastry	425	220
8	very hot – bread	450	230
9	rarely used – bread	475	250

Things to do

1 Draw or copy the diagram of a gas or electric cooker. Label the hob, grill and oven. On an electric cooker, label the hot plates, on a gas cooker, the burners.
2 Find a picture of a spit, a split level cooker, and a solid fuel cooker. Draw them or stick them in your book, label clearly.
3 Make your own oven temperature chart. Either copy the chart into your book, or make a wheel. For the wheel you will need
two circles of card approximately 10 cm diameter, scissors, clip, felt pens in two colours, ruler.
   Cut one circle of card 2cm smaller than the other.

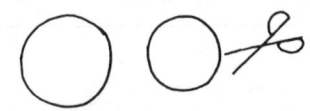

Divide the larger one into eight equal segments. Cut a triangular 'window' in the smaller one, the size of one of the segments.

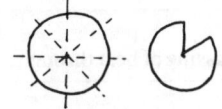

Number the segments 2-9 around the edge of the larger circle. Clip the two circles together. Turn the inner circle so that the 'window' shows the different segments, and fill them in with the equivalent temperature in electricity.

4 Look at this list of cooking methods. Copy down the sentences, filling in the correct cooking method for each part of the cooker.
   frying, baking, steaming, boiling, stewing, poaching, roasting, grilling, toasting
   a) The hob is used for f——, b——, s——, p——, s——.
   b) The grill is used for g——, t——.
   c) The oven is used for ba——, r——, s——.
5 Choose oven temperatures in °C and gas nos. for cooking these foods:
   flapjack, soda bread, yorkshire pudding, beef stew, meringues
6 Name four pieces of equipment in the home that are controlled by a thermostat.

Boiling

Water boils when it reaches a temperature of 100°C. The liquid evaporates and turns into steam. If you continue to boil an uncovered pan of liquid it will boil dry, that is, all the liquid will evaporate into steam. Boiling is cooking in liquid at 100°C.

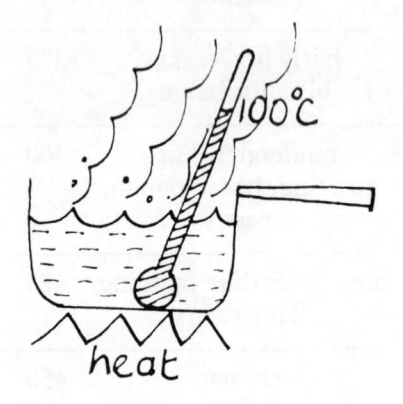

Conduction

The pan on the hob is heated by the ring or burner. The metal of the pan becomes hot and carries the heat by **conduction** to the contents which become hot and boil at 100°C. Pots and pans for cooking must be good conductors of heat, to pass the heat to the food inside and to save fuel. Most metals are good conductors of heat.

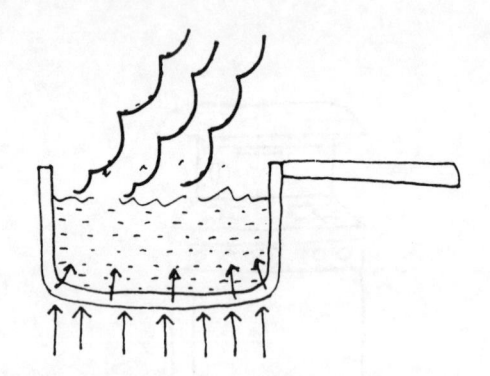

Things to do

1. Try this experiment.
   Half fill a saucepan with cold water. Place a wooden spoon and a metal spoon in the pan. Heat the water until it boils. Hold one spoon in each hand.
   What do you feel?
   Copy the sentences into your book, filling in the missing words.
   a) The metal spoon becomes ---- because it carries or -------- heat along it.
   b) The wooden spoon feels ----.
   c) The metal spoon is a ---- conductor of heat.
   d) The wooden spoon is a ---- conductor of heat.

2. What is your saucepan made from? Is it a good conductor of heat?
   What is the handle made from? Is it a good conductor of heat?

3. Explain why a wooden spoon should be used to stir a hot mixture in a saucepan.

4. Choose four materials from this list that are good conductors of heat:
   rubber, plastic, steel, paper, copper, glass, aluminium, wood, iron

5. Choose the correct words from the list below to complete the instructions for boiling an egg.
   a) Place egg in a saucepan. Cover with ----, add ----.
   b) Bring to the boil. Turn heat to ----.
   c) The pan becomes ----. It is made from ---- which is a ---- conductor of heat.
   d) The water boils at ---- and ---- into ----.
   lid, steam, high, low, good, 100°C, hot, sugar, water, salt, smoke, poor, metal, evaporates

6. Fill in the chart with the correct times for boiling an egg.

	Minutes cooked in boiling water
soft	
medium	
hard	

What is a coddled egg?

> *New word*
> **conduction** – the passing of heat through a solid object

22

Stuffed Eggs

Ingredients
(for 4)
4 eggs (size 2–3)
1 tsp salt
75 g grated cheese
1 tblsp salad cream or mayonnaise

Utensils
saucepan, wooden spoon, basin, fork, colander, teaspoon

Method

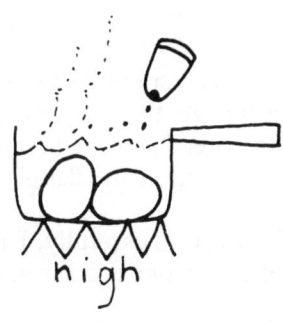

1 Place eggs in saucepan, cover with cold water, add salt. Place on high heat until water boils. This is seen when water steams and bubbles.

2 Turn the heat off. Cover the eggs with a lid and leave to cook in the heat of the water for 15 minutes.

3 Drain off the water in a colander or sieve. Tap eggs lightly to crack the shells and make them easier to remove.

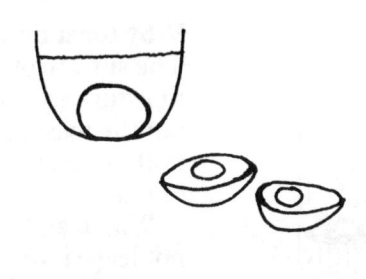

4 Place eggs in cold water at once to prevent them becoming black. Remove the shells, throw away. Cut eggs in half.

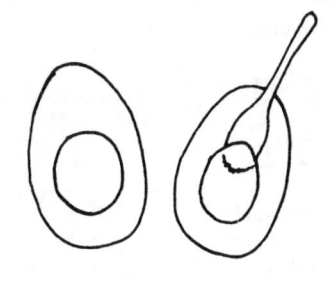

5 Remove yolk carefully using a teaspoon.

6 Mix yolk, cheese and salad cream in a basin to a stiff paste.

7 Refill each egg half with the mixture using two teaspoons.

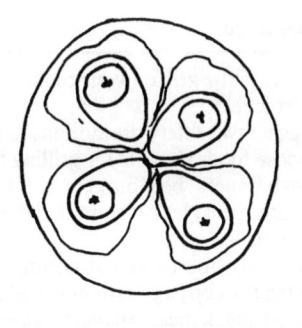

Decorate the top with parsley or paprika pepper. Serve with brown bread and butter and a salad.

Grilling

The grill may be either at eye level or waist level. The heat comes from the flames or elements which throw the heat directly on to the food underneath. Only the top surface is cooked and the food must be turned regularly so that both sides cook and the heat cooks the centre of the food. This direct heat is called **radiation**. The heat of the sun on a hot day and the warmth from the bars of an electric fire are produced by radiation.

Remember that the radiant heat of the grill could burn your fingers and that the food will be hot, so use tongs or a palette knife to turn it. Grilling is a fast, dry method of cooking thin pieces of food, e.g. bacon, sausages, fish, steak, hamburgers, chops. Foods containing a lot of fat (chops, sausages) will spit. The fat will reach the flames and could cause a fire. Turn the heat to low when cooking these foods.

Why toast curls

The surface of the bread nearest the heat loses moisture as it becomes hot. As the surface dries out the toast curls. If the bread is turned often, both sides dry out evenly resulting in a flat piece of toast.

Watch all food under the grill carefully. Do not leave it unattended as it cooks quickly and may burn easily.

Things to do

1 Where is the grill on your cooker? Is it at eye level or waist level?
 Suggest two safety precautions to take when grilling.
2 Choose foods suitable for grilling from this list:
 fish fingers, beef burgers, fish fillet, chips, egg, tomato, peas, leg of lamb, bread, mushroom, whole chicken
3 When a recipe asks for seasonings, what will you add?
4 Welsh Rarebit, a glass of milk and an apple would make a good snack meal. Suggest three other snacks that could be made using the grill and the accompaniments that you would serve with them.
5 What other piece of equipment besides a grill can be used to make toast? Make a slice of toast by each method. Which is quicker? Which method do you prefer? Why?
6 Fill in the crossword.

Clues down
1 Edge of a slice of bread
2 Grilled food becomes this colour
5 Grilling is a —— method of cooking.

Clues across
2 Grilled food may do this easily
3 Grilled bread
4 Type of heat from the grill

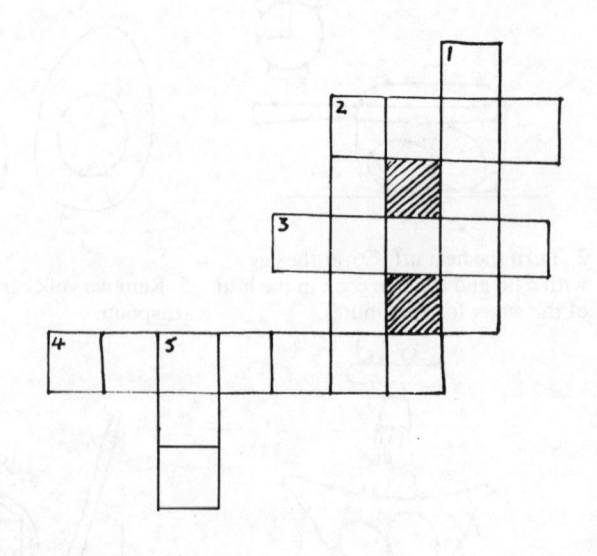

New word
Radiation – heat rays falling directly on an object

Welsh rarebit

Ingredients
(for 4 people)
150 g Cheddar cheese
25 g butter
40 ml milk
4 slices bread
¼ tsp salt
¼ tsp pepper } seasoning
¼ tsp dry mustard
parsley and tomato to garnish

Utensils
grater, basin, teaspoon, wooden spoon, measuring jug, palette knife, serving plate, grill pan

Method

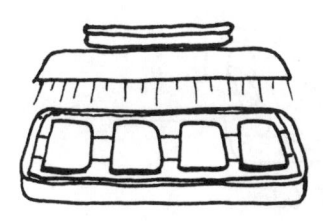

1 Turn the grill on full. Adjust the grid to near the heat. Put serving plates to warm. Toast bread lightly on both sides, turning often to keep flat. Put toast to keep warm on plates while cheese mixture is prepared.

2 Grate the cheese on the large holes of the grater.

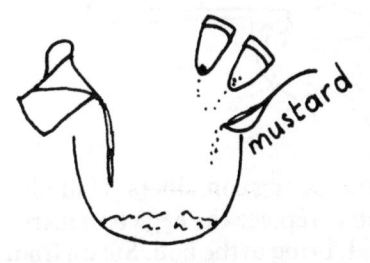

3 Place cheese in basin. Add seasoning and milk. Mix to a thick paste.

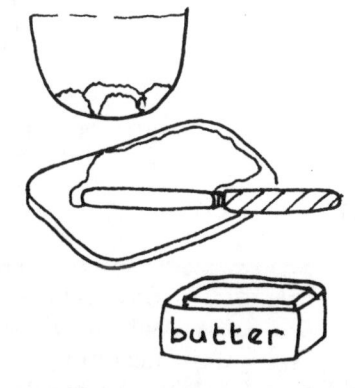

4 Spread toast with butter. Divide cheese mixture into 4, spread each slice evenly with the cheese mixture, taking it up to the edges to protect them from burning.

5 Place under the grill until brown. This happens quickly, so stay and watch it cooking.

Serve Welsh rarebit on warm plates garnished with parsley and a slice of tomato. It may be eaten as a snack lunch or supper on its own, or served with soup and fruit for a more filling meal.

Add a poached egg on top of the rarebit to make a buck rarebit.

Other toppings include grilled bacon, fried mushrooms and thinly sliced apple.

New word
Seasoning – flavouring added to improve the taste of food, usually in small amounts, e.g. salt, pepper, herbs and spices

Poaching

Poaching is cooking in hot liquid just below boiling point. When water boils at 100°C, bubbles rise quickly and steam is given off. During poaching, the liquid does not boil although steam is given off.

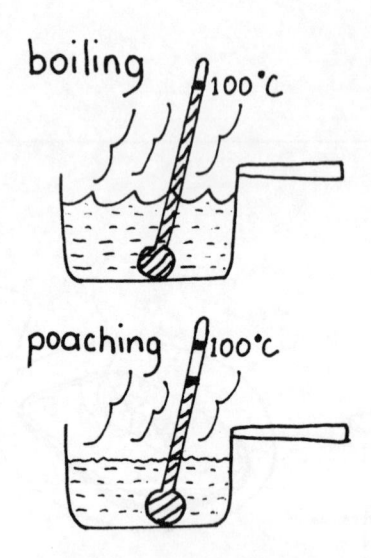

To check the temperature of a liquid for poaching, bring it to the boil, then lower the heat and wait until it just stops bubbling.

Poaching is a gentler method of cooking than boiling and is suitable for many foods, e.g. eggs, fish, (smoked haddock, kippers) chicken and some other meats.

The liquid used for poaching may be milk, water or stock. Fish may be poached in milk, then the milk used to make a sauce to serve with it. Chicken may be poached in stock, and eggs are usually poached in water.

Poaching an egg
A frying pan or saucepan may be used. Add enough water to half fill the pan. Add salt and vinegar; these give flavour to the egg and also help it to set quickly. Bring the water to the boil, lower the heat and wait until the bubbles die down. Break the egg into a cup. Using a wooden spoon, stir the water in the pan vigorously in a circle to make a small whirlpool in the pan, drop in the egg.

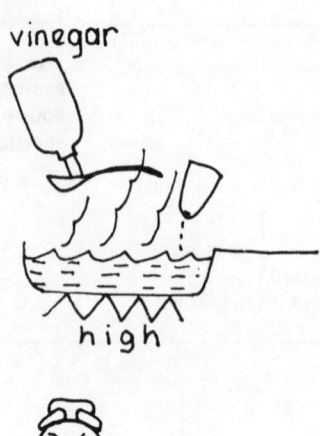

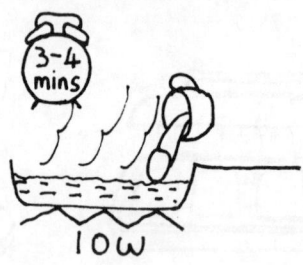

As the egg hits the hot water, the heat sets it, the salt and vinegar also help this so the egg forms a neat lump. Poach for 3–4 minutes.

A special pan called an egg poacher can also be used. This produces a quite different egg, really a steamed egg.

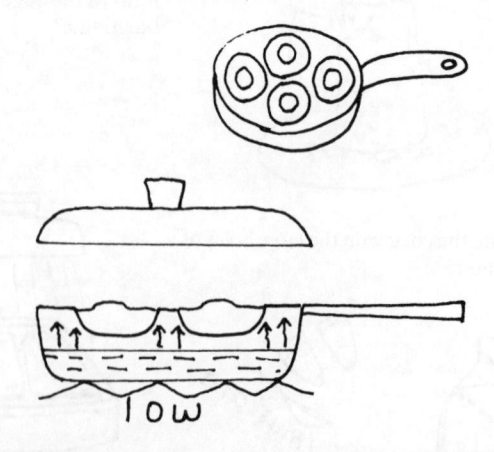

Butter and season the egg containers. Half fill the pan with water, replace the egg containers. Cover with the lid, bring to the boil. Steam from the boiling water underneath the eggs rises to cook them.

Egg florentine

Ingredients
(for 4 people)
4 eggs
1 packet whole spinach leaves (frozen about 200 g)
50 g Cheddar cheese
25 g butter
1 tblsp vinegar
salt, pepper

Utensils
saucepan and lid, grater, colander, wooden spoon, teaspoon, tablespoon, draining spoon, flat ovenproof serving dish (e.g. lid of casserole dish)

Method
Collect ingredients and utensils.

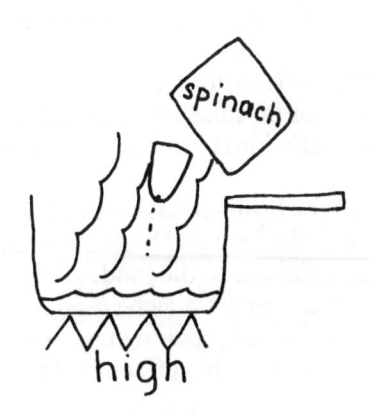

1 Cook the spinach: place 125 ml water in a saucepan, add ½ tsp salt. Bring to the boil, add the frozen spinach.

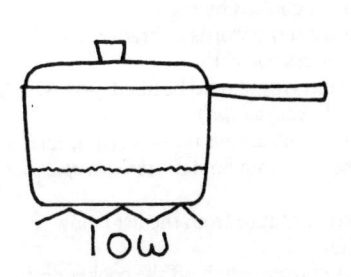

2 Cover and simmer on low heat for 15 minutes or until the spinach is soft.

3 Grate the cheese. Place serving dish to warm under low grill.

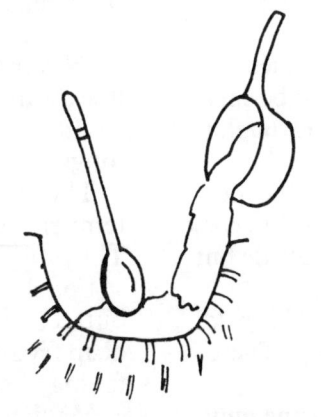

4 Drain the spinach in a colander pressing out the water with a wooden spoon. Melt the butter in the saucepan, add the spinach and mix well.

5 Spread spinach on ovenproof dish, leave to keep warm.

6 Rinse saucepan and re-use to poach the eggs. Half fill with water, add salt and vinegar. Poach the eggs for 3 minutes. Lift out and drain.

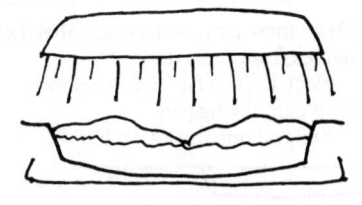

7 Lay the eggs on the spinach. Cover evenly with the grated cheese, brown under a hot grill.

Serve at once with toast or bread for lunch or supper.

Frying

Frying is a quick method of cooking in hot fat. The outside of the food is cooked and browned by contact with the hot fat. This browning gives the food a pleasant flavour and appetising colour. Heat from the hot fat then passes into the food, cooking it right through.

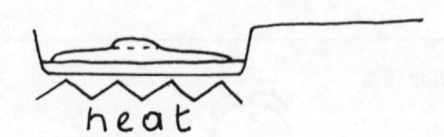

Food must be turned so that both sides brown evenly. Foods to be fried must be small and thin so that heat passes quickly through to the centre.

If the fat is very hot, the outside of the food will burn before heat reaches the inside to cook it, e.g. a 'rare' steak is well cooked outside but almost raw inside.

If the fat is not hot enough, the food is pale and greasy instead of brown and crisp. There are three methods of frying:

a) Shallow frying – in a frying pan using only enough oil to cover the pan thinly, e.g. eggs, hamburgers, fish, fish fingers.

b) Deep frying – in a deep fat frying pan or chip pan with a basket. This contains enough fat to cover the food (about half full), e.g. chips. An electric thermostatically controlled frier may be used.

c) Dry frying – the fat contained in fatty foods is used to fry them, no extra fat is needed, e.g. sausages, bacon, fatty chops.

The fat used for frying is hotter than boiling water. For this reason care must be taken when frying.

a) Never leave a frying pan or deep fat pan unattended; hot fat can catch fire.

b) Keep the heat under the pan medium to low. Allow fat to heat up gently.

c) Never add wet foods to hot fat. Dry them first, water makes hot fat splash dangerously.

d) Lower foods gently into fat using spoons or tongs.

The fats chosen for frying must reach a high temperature to cook the food and brown it well. They may also give flavour to the food.

For shallow frying, any fat may be used: butter, margarine, oil, lard, cooking fat. For deep frying, only lard, cooking oil, or cooking fat may be used.

Many foods that are suitable for frying are also suitable for grilling, e.g. bacon. When fried, they usually contain more fat.

Things to do

1 Draw these pans for frying. Label them with the correct names from this list:
   electric deep fat frier, omelette pan, deep fat pan or chip pan, frying pan
   Suggest one use for each.

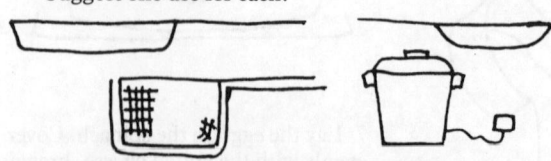

2 Are these sentences true or false? Copy the true ones, change the false ones to make them correct.
   a) Frying is a quick method of cooking in hot fat.
   b) All foods can be fried.
   c) Foods to be fried must be large and tough.
   d) In shallow frying, the food is covered in fat.
   e) Chips, Scotch eggs and fish are deep fried.
   f) Sausages are deep fried.
   g) Wet foods added to hot fat will make the fat 'spit'.
   h) Butter can be used for deep frying.
   i) Hot fat can easily catch fire.

3 Design a poster to show the three methods of frying and the foods that can be cooked by each one.

4 Choose one word from the words in brackets to complete the sentences correctly.
   a) Pans for deep fat frying should be filled ⎯⎯ with fat or oil. (*full, half full, quarter full*)
   b) Food to be fried must always be ⎯⎯. (*wet, cold, dry*)
   c) Food should be ⎯⎯ into hot fat. (*thrown, splashed, lowered*)

   Choose the correct phrase from the list below to complete the sentence:
   If a fat pan catches fire, switch off the cooker and ⎯⎯.
   move the pan, cover with a lid, throw cold water on it, cover with a dry tea towel, cover with a wet tea towel

Fried sandwiches

Ingredients
(for 2–3 people)
6 slices white bread
25 g butter
3 slices cooked ham
3 slices cheese, about 35 g
1 egg
150 ml milk
salt, pepper
oil for frying

Utensils
frying pan, tongs, basin, fork, palette knife, kitchen paper, cook's knife, plate, serving plate

Method
Collect ingredients and utensils. Leave serving plates to warm.

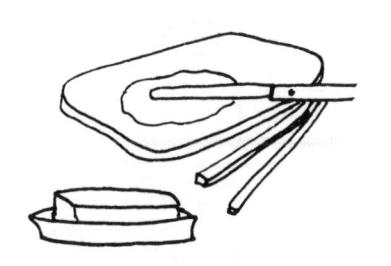

1 Remove crusts from bread, spread with butter.

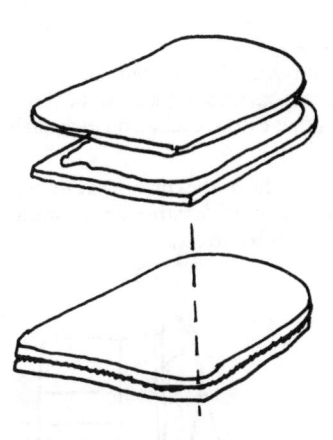

2 Lay slices of cheese on the buttered side, cover with slice of ham. Top with a slice of bread to make a sandwich. Cut into 2.

3 Beat egg, milk and seasoning in a basin with a fork.

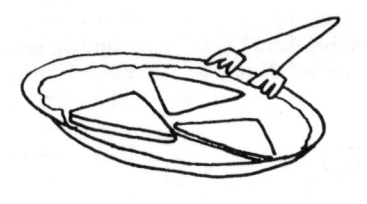

4 Pour egg mixture on to a plate. Dip the sandwiches in this, turning them to absorb the mixture.

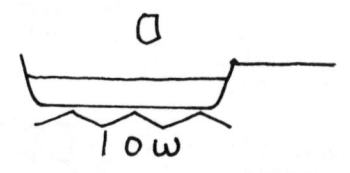

5 Pour enough oil into a frying pan to come half way up the sides. Place on a low heat. Test the heat with a bread crust: when it sizzles and browns in 1 minute, the fat is hot enough. *Do not overheat the oil.*

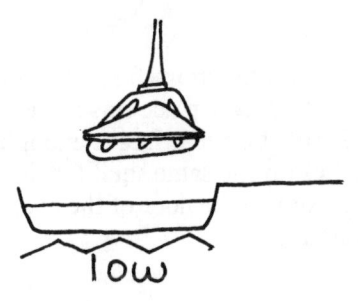

6 Lower the sandwiches gently into the hot fat, cook for 5 minutes on each side until crisp and golden.

Drain the cooked sandwiches on kitchen paper to remove excess fat. Serve at once with a crisp salad or celery sticks for lunch or supper. (Suggested menu: fried sandwiches, green salad, fruit fool, biscuit, coffee)

Cooking in the oven

Food may be cooked in the oven by:
a) baking, e.g. biscuits, cakes, bread
b) roasting, e.g. meat, potatoes
c) stewing, e.g. meat, fruit.

In each case the air inside the oven is heated by the flames at the back of a gas oven, or the electric elements at the side of an electric oven. The hot air rises to the top of the oven, cooler air moves to the bottom where it is heated and rises again. This produces a circular movement of air called **convection**, which helps to cook the food.

The air must move freely to cook the food evenly. Never place tins or dishes against the sides of the oven. Leave enough space between dishes on the same shelf for the air to circulate. Do not use the floor of the oven for baking or roasting.

The container in which the food is placed will become hot and pass heat to the food inside it by conduction.

The metal sides of the oven also become hot and help to cook the food by radiation.

There are three areas or zones of heat in a gas oven. The number on the gas knob shows the heat at the centre of the oven. Because hot air rises by convection, the top shelf is about 10° hotter than this and the bottom shelf 10° cooler. In an electric oven the zones of heat are not so marked.

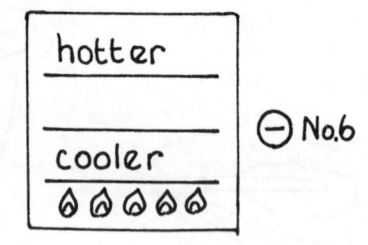

Things to do

1 Experiment to find the zones of heat in a gas oven.
   Light the gas oven at gas no. 6. Arrange the shelves in position. Collect three baking trays. Cut a piece of white paper into three even-sized rectangles. Number them 1, 2, 3. When the oven is hot, place 1 on the top shelf, 2 on the centre shelf and 3 on the bottom shelf.

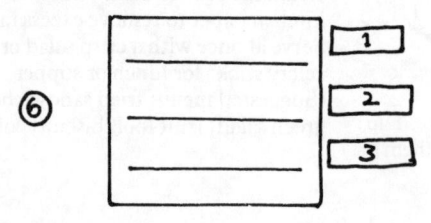

Leave for 15 minutes. Remove from oven and switch off. Look at the pieces of paper, then copy this diagram into your book, sticking the paper samples in position.

The paper from the top shelf is brownest.
The top shelf is the hottest because hot air ------ due to c---------.

2 Copy the sentences, fill in the missing words to show the methods of cooking in the oven.
   The oven may be used for
   a) ------, e.g. scones ----.
   b) r------, e.g. ----, ----.
   c) s------, e.g. ----, ----.

3 Design a poster to show six rules for oven safety.

4 The oven in the diagram opposite is set at no. 6. Write or draw suitable dishes that can be cooked on each shelf at this temperature.

5 Which of these diagrams shows the correct way to stand when opening the oven door? Copy the correct one into your book, explain why it is correct.

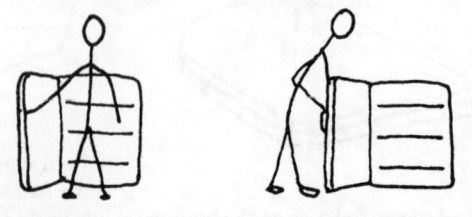

New word
convection – the rise and fall of hot air in a circular movement

Flapjack

Ingredients
125 g rolled oats
75 g brown sugar
75 g margarine
2 tblsp golden syrup

Utensils
baking tin measuring 200 mm square or 170 × 220 mm, saucepan, wooden spoon, tablespoon, basin

Method
Light the oven at elec. 180°C (gas no. 4).
Collect the ingredients and utensils.

1 Grease the baking tin.

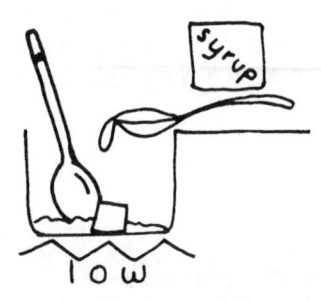

2 Place margarine, golden syrup and sugar in a saucepan. Stir over a low heat using a wooden spoon until the fat and sugar dissolve and the mixture becomes clear. Do not boil.

3 Remove from heat. Stir in the oats. Stir thoroughly to form a thick mixture.

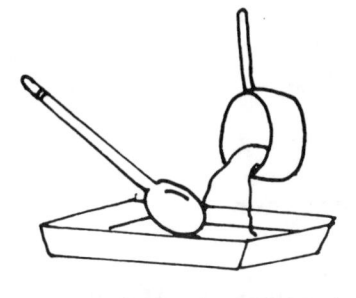

4 Press the flapjack into the tin using the back of the spoon to flatten the surface.

5 Bake for 20 minutes or until light golden brown.

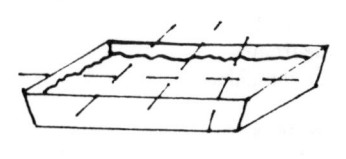

6 Remove from the oven and while still warm, cut the mixture into 8 even-sized squares, cutting first down the centre.

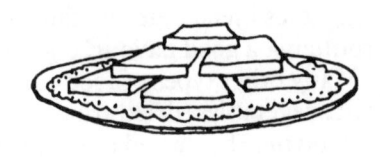

Leave the biscuits in the tin until they become cold. They are too soft to move while still warm. Serve flapjacks for tea or morning break.

Stewing

Stewing is the cooking of food in hot liquid in a covered container. Food can be stewed either on the hob or in the oven. The food and cooking liquid are usually served together. The liquid must not boil, but is kept simmering just below boiling.

This method is useful for tough pieces of meat. The long slow cooking in liquid helps to soften them. The flavours of the meat and vegetables blend together during cooking, producing a good flavour, e.g. beef stew.

Many other types of food can be stewed, e.g. fruit, vegetables, dried fruit.

Food may be stewed in a covered saucepan or a casserole dish, from which the finished result may take its name, e.g. chicken casserole. A casserole dish must be made of a material that will not crack in the oven, i.e. it must be **ovenproof**. It must be covered, either with a lid or aluminium foil, to prevent loss of moisture and flavour. Casserole dishes may be made from many materials – earthenware, cast iron, enamelled steel, ovenproof or toughened glass (this may be clear or coloured), stainless steel.

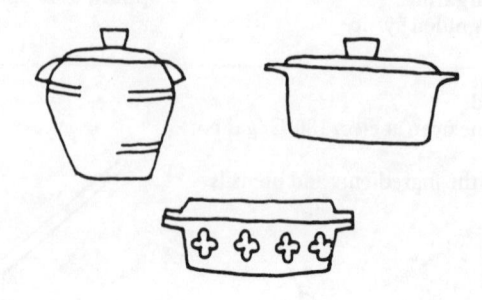

Handles and knobs on these dishes must also be ovenproof.

A thermostatically controlled electric stewing pan usually made of earthenware can be used. The food cooks inside it for several hours at a low temperature; an oven is not needed.

Things to do

1 Choose the foods from this list that are suitable for stewing:
   apples, chicken, eggs, peas, hamburgers, potatoes, celery, bacon, ham, fish, mince, pork chops, tomatoes
2 Add vegetables and a sweet to the following dishes to make a complete meal:
   a) sausage hot pot b) beef stew c) minced beef casserole d) chicken casserole
3 Choose the materials from this list that would make ovenproof dishes:
   earthenware, tin, clay, glass, copper, toughened glass, stainless steel, wood, aluminium, cast iron
4 Stews and casseroles are popular all over the world. Which countries do the following stews come from?
   coq au vin, Lancashire hot pot, osso buco, ratatouille, goulash, cawl.
   Find the names of three traditional British stews.
5 Fill in the crossword.

Clues down
1 To cook in liquid at a low temperature.
3 See 4 across.
5 Food may be cooked in a casserole in this part of the cooker.
6 Stews may be cooked on this part of the cooker.

Clues across
2 When cooking a stew, the liquid must do this.
4 And 3 down. Food that is improved by slow cooking.
7 Stews and casseroles must never do this.
8 A metal used for some casserole dishes.

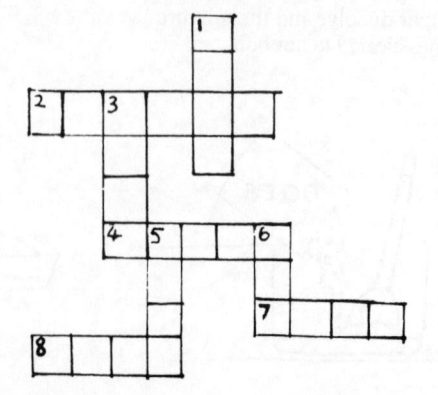

New word
ovenproof – describes something that will not be damaged by normal use in an oven

Quick sausage hot-pot

Ingredients
(for 2–3 people)
200 g sausages (beef, or pork and beef)
1 small onion (about 100 g)
1 carrot
1 stick of celery
1–2 potatoes (400 g)
250 ml hot water
1 small tin baked beans (220 g)
1 tblsp cooking oil

Utensils
frying pan, tongs, 500–600 ml ovenproof dish, plate, kitchen paper, vegetable knife, peeler, chopping board, tin opener, measuring jug

Method
Light the oven at elec. 200°C (gas no. 6). Collect all ingredients and utensils.

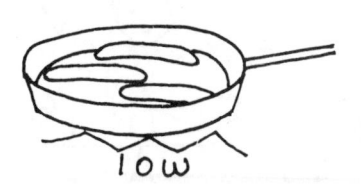

1 Fry the sausages in a frying pan over medium to low heat, until they start to brown, (about 10 minutes). No extra fat is needed as sausages contain enough of their own. Keep the heat steady to prevent them sticking.

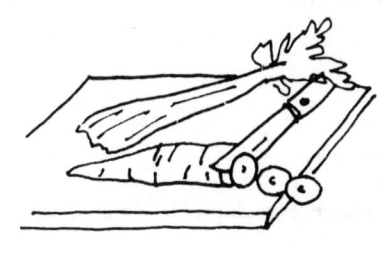

2 Meanwhile, prepare the vegetables using a peeler, vegetable knife and a chopping board. Peel and slice the onion and carrot. Wash and slice the celery. Peel and thinly slice the potato. Wrap up the peelings and throw them away.

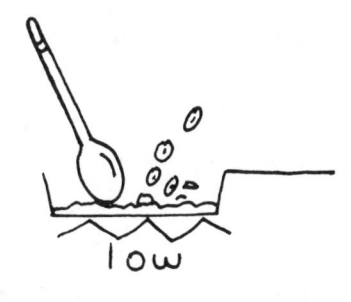

3 Remove sausages, drain on kitchen paper. Add 1 tblsp oil to the frying pan if needed. Fry the vegetables for 5 minutes, stirring with a wooden spoon.

4 Open the tin of beans. Measure the water.

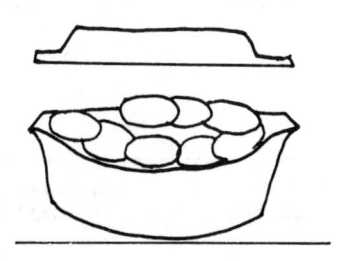

5 Place sausages, fried vegetables, water and baked beans in the casserole dish. Arrange the thin slices of potato overlapping each other to cover the top. Cover with a lid or foil. Remove this 10 minutes before the end of the cooking time to brown the top.

6 Bake for 30 minutes, or longer if possible, until the potatoes are cooked and golden brown.

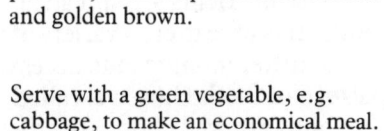

Serve with a green vegetable, e.g. cabbage, to make an economical meal. (Suggested menu: sausage hot-pot, cabbage, apple crumble, custard, water).

3 The food we eat

The food we eat is needed by the body for: growth, energy, protection, repair.

Food contains a variety of nutrients which help the body to work, and which provide energy. The study of food and how it is used by the body is called **nutrition**.

The names of the most important nutrients are:
a) protein, which provides material for growth, maintenance and repair
b) carbohydrate, which provides energy
c) fat, which also provides energy
d) micronutrients, the small amounts of vitamins and minerals that provide for protection and repair.

These nutrients are spread among many foods. It is wise to eat a variety of food to obtain all the different nutrients necessary. The pattern of foods chosen is called the diet, and if the foods are well chosen the diet is well balanced, i.e. it contains the correct amount of nutrients for all the body's needs. If the diet is not well balanced, discomfort and illness can occur.

too thin, too fat, tooth decay, poor skin

A well-balanced meal contains a variety of foods. It would be unwise to eat a meal of only four apples, or four jars of yoghurt, but included in a meal together they could balance well, e.g. cheese sandwich, apple, yoghurt.

Check that all meals and snacks contain some protein food, some food for energy, fresh fruit and/or vegetables, a drink and a food containing fibre.

Check that meals and snacks do not contain too much sugar, too many sweet, sticky foods, or fizzy drinks. Avoid eating sweets and sweet snacks in between meals; choose fruit, nuts or raw vegetables instead.

Things to do

1 Design a poster to advertise eating a well-balanced diet.
2 Rearrange the letters in the words to give the names of foods:
   a) gegs, tmea, sfih, spluse, hecees, klim, rcleeas (all contain protein)
   b) sfta, gsura, kcae, mja, rceeasl, tptoao, rdbea (all provide energy)
   c) klim, rgonea, bbcagae, tmea, lfruo (all contain micronutrients)
   d) rftui, nbaes, vgteblseea (all provide fibre)
3 Are these meals well balanced? If not, suggest how they could be altered to make them well balanced.
   a) tomato soup, cornish pasty, chips, baked beans, apple pie, cream
   b) fried bacon, sausage, egg, chips, bread and butter, ice-cream, tea
   c) crisps, fizzy drink, bar of chocolate, paste sandwich
   d) salad, fruit squash, cake
4 Write out the meals you have eaten in one day. Check that they are well balanced. If not, how could they be improved?

New word
nutrition – the study of food and how it is used by the body

Flavour and texture of food

We must eat to live, but certain foods are eaten because we like them more than others. Why do we like some foods and dislike others? This depends partly on the flavour and texture of the food.

Flavour
Food is tasted in the mouth with the tongue. The tongue can detect these flavours only in the areas marked:
a) sweet b) salty c) sour d) bitter.

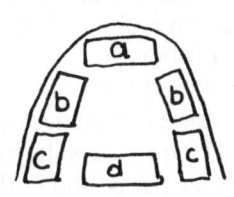

There is an important difference between sour foods and bitter foods: a lemon is sour and a grapefruit is bitter.

The sense of taste is also affected by colour, appearance, smell and texture. To be appetising, food should look attractive and smell pleasant. Meals should be chosen to include a variety of textures and colours.

Foods are expected to be certain colours, e.g. a banana is yellow, therefore a yellow jelly could be banana flavoured, but it could also be pineapple or lemon flavoured. It would be a surprise if it tasted of strawberry or orange. For this reason many foods have artificial colour added to them which may alter or improve the colour of the food. Some foods have preservative added to preserve or keep the colour during storage. A good cook will think about the colour of food in a meal and try to keep the colour of foods during cooking.

Texture
This describes how food feels in the mouth. The texture of food can be described in many ways: soft, hard, crisp, crunchy, smooth, lumpy, runny, spongy, rubbery, tough, etc.

The textures of foods in a meal should also be considered. A meal containing only smooth or runny foods would be dull to eat.

Things to do

1 Name two foods that taste
a) sweet b) salty c) sour d) bitter
2 Make a list of words to describe the texture of foods. Choose words from the list to describe the texture of the following foods:
   celery, biscuit, fried egg, scrambled egg, apple, banana, fruit fool, flapjack, mashed potato, toast, bread.
3 Choose a food you enjoy. Explain why you like it describing the flavour and texture.
4 Choose a food you dislike. Explain why you dislike it describing the flavour and texture.
5 Colour is added to many foods because they are expected to be certain colours. Look at the labels on these foods. How many of them have added colour?
   butter, margarine, custard powder, blancmange, yoghurt, orangeade, orange juice, tomato ketchup, tinned peas, fruit jellies, cola drinks, 'instant' fruit puddings
6 Imagine eating these meals. What is wrong with them? Alter one or more foods in each meal to make it more attractive.
a) Baked fish, parsley sauce, mashed potato, cauliflower, rice pudding, pears
b) Macaroni cheese, strawberry fruit fool, lemon squash
c) Beef stew, baked jacket potato, chocolate pudding, chocolate sauce
d) Fried fish, chips, beans, apple fritters
7 Taste and colour experiment
Make up a yellow jelly. Divide into two halves. Add green colour to one half. Leave both to set.
Make up an orange jelly, divide into two halves. Add red colour to one half, leave to set.
Ask three or more friends to taste the four jellies and guess the flavours. Copy and complete the following table.

	1	2	3
yellow			
green			
red			
orange			

How many people guessed the correct flavour of all the jellies?
Does the colour of the jelly affect the flavour it is expected to be?

Meat

Meat is the flesh of animals. The animals most commonly used for meat in this country are:
a) cows to produce beef b) sheep to produce lamb c) pigs to produce pork and bacon. Other animals are also used, e.g. rabbit, deer.

Animals are killed and butchered into different cuts which are often named after the part of the animal from which they come, e.g. leg of lamb, shin of beef, belly of pork.

Some inside organs of animals can be eaten. These are called offal, e.g. liver, kidneys, heart, tripe from the inside lining of the stomach.

The flesh of birds is also used, e.g. chicken, duck, turkey, goose. These are known as poultry. Some birds (pheasant, partridge) are bred for sport and are known as game birds.

Meat is useful, though not essential in the diet. It contains an important nutrient for body growth, maintenance and repair, called **protein**. It also contains other useful nutrients: fat for energy, iron for healthy blood, other micronutrients for protection and maintenance.

Protein

The body needs about 60 g protein daily. This amount would be provided in 200 g meat. But meat is an expensive food, so not all the day's protein comes from meat. Other foods also contain protein and should be eaten with or instead of meat, e.g. fish, milk, eggs, cheese, nuts, cereals, pulses. To make meat go further, to make it more filling and to add variety, it is eaten with potatoes and vegetables as in the traditional British 'meat and two veg.', or with rice, pasta, pastry, and other filling foods.

Some pieces of meat are cheaper than others. Cheap joints or cuts are usually tougher and can be made tender by long slow cooking. The more expensive cuts are tender and suitable for quick cooking by grilling or frying.

Mince is fairly cheap, and good value if carefully chosen. Butcher's mince will contain up to 40% fat so choose meat that is bright red in colour and does not look too fatty. Use the mince the day it is bought.

Meat may be bought in a butcher's shop or a supermarket, where it is often ready weighed and wrapped. The butcher's shop must be clean and show the price per pound of the meat clearly. The supermarket must also show the price per pound, the weight of the cut, and the cost of that particular piece.

Meat is used in many meat products, e.g. pies, patés, beefburgers, sausages, meat balls. It is mixed with other ingredients but the law states that a certain amount of meat must be included, e.g. pork sausages must contain at least 80% pork, canned pie fillings at least 35% meat, meat pies at least 25% meat.

A meat substitute called T.V.P. made from soya beans is used to replace meat or make it go further in many dishes.

Things to do

1. Fill in the missing word:
    a) Meat is the flesh of ————.
    b) Lamb comes from ————.
    c) Beef comes from ————.
    d) Pork comes from ————.
    e) Bacon comes from ————.
    f) Two kinds of poultry are t————, ————.
2. Make a list of foods containing plenty of protein. Design a poster to show these foods.
3. Read these sentences carefully, alter *one word* in each and rewrite correctly.
    a) Meat is a poor source of protein.
    b) Protein is needed for growth and health.
    c) Meat is always essential in the diet.
    d) Eggs, milk, butter also contain protein.
    e) Meat contains copper for healthy blood.
    f) Tough meat is good when grilled.
4. Make a list of dishes that can be made with mince.
5. Visit a butcher's shop and a supermarket to find out the prices of different cuts of meat.
6. Work out the cost of the following:
    a) ½ lb mince at 94p/lb
    b) 6 oz mince at 85p/lb
    c) 3 lb leg of lamb at £1.25/lb
    d) 4 oz liver at 40p/lb

> *New word*
> **protein** – a nutrient found in a wide variety of foods that aids the body in growth, maintenance and repair

Meat loaf

Ingredients
(for 3–4 people)
300 g mince
75 g rolled oats
1 egg
½ tsp mixed herbs
salt, pepper
1 tsp Worcester sauce
6 tblsp cold water
1 onion

Utensils
a 1 lb loaf tin, grater, fork, teaspoon, tablespoon, aluminium foil

Method
Collect ingredients and utensils. Light the oven at 200°C (gas no. 6). Grease the tin.

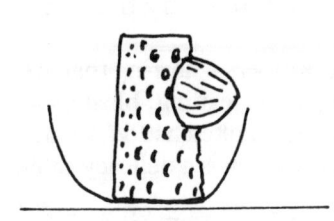

1 Peel the onion. Grate it into a mixing bowl.

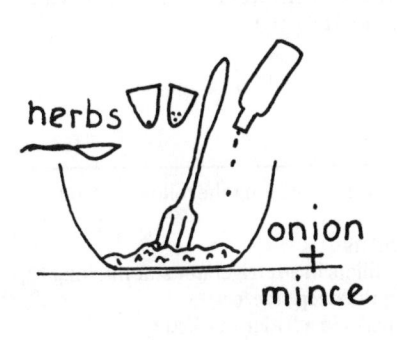

2 Add the mince, breaking it up with a fork. Use a fork to stir in all the ingredients, mixing thoroughly.

3 Press the meat mixture into the tin firmly, using the back of a spoon.

4 Cover the tin with aluminium foil, bake for 30–35 minutes. Test with a fork; if the juices run clear it is cooked. Leave in the tin for 1 minute before turning out.

Meat loaf may be served hot for a main meal, or cold for parties and picnics. (Suggested menus: hot meat loaf with baked jacket potato, home-made tomato sauce, broccoli or cabbage; cold meat loaf with potato salad or baked potato, tomato and onion salad).

37

Fish

Fish are caught in the sea around the coast, as well as inland in rivers, lakes and reservoirs. The wide range of different types of fish add colour and variety to the diet.

All fish contain protein in their flesh and are a useful protein food. They are divided into three groups according to their appearance and the other nutrients they contain.

a) White fish e.g. cod, plaice, haddock, coley, skate. The flesh is white though the skin may be brown, grey or coloured. The flesh contains protein and useful micronutrients – iodine, fluorine, calcium, phosphorus, Vitamin B. White fish do not contain fat, this makes them easier to digest, though fat is often added during cooking.

b) Oily fish e.g. herring, mackerel, trout, salmon, sardines, tuna. The flesh is coloured: mackerel and herring are greyish, salmon is pale pink when cooked. As well as protein, the flesh contains oil or fat. In addition to the micronutrients found in white fish, the oil contains Vitamins A and D.

c) Shell fish e.g. prawns, scampi, shrimps, cockles, mussels, crab, lobster. As the name suggests, they are inside a hard shell. The flesh contains protein and micronutrients, but little fat.

Choosing and buying fish

a) Fresh fish bought from a fishmonger. The shop or stall must be clean and the fish displayed chilled on ice to keep fresh. It should be firm and bright-eyed and both shop and fish should smell clean. The fishmonger will clean and prepare the fish, it may be left whole or cut into:

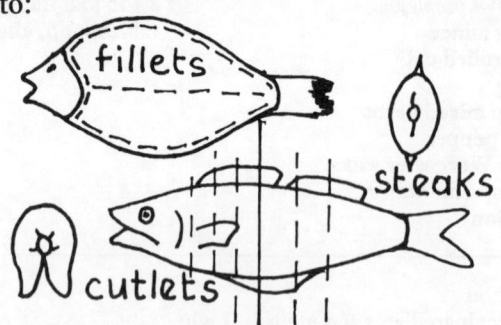

fillets, steaks, cutlets

Eat fresh fish the day it is bought. Cover lightly and store in the refrigerator until needed.

b) Frozen fish is a useful convenience food. The fish is deep frozen as soon as it is caught at sea. It is cleaned and prepared and ready for use. Store according to the directions and 'star markings' on your refrigerator ice-box, or in a freezer for up to three months.

c) Smoked fish, e.g. kippers – herring that are smoked after gutting and cleaning, bloaters – whole smoked herring. Smoked haddock is also popular and smoked salmon is a delicacy eaten raw.

d) Tinned fish, e.g. salmon, sardines, tuna, pilchards. These are useful to keep in store for use in salads, sandwiches, fish cakes and pies.

e) Fish products e.g. fish cakes, fish fingers, fish pastes and spreads. Other ingredients are added but they must contain a certain amount of fish, e.g. fish cakes must contain at least 35% fish and fish pastes at least 70% fish.

Things to do

1. The letters in the names of these fish have been mixed up. Rearrange them correctly to give the names of
   a) oily fish: noslam; ntua; stpar; rrhgnie
   b) white fish: eplcia; ycloe; khea; elos, lbhiaut
   c) shell fish: rabc, styore, lkwhe, nwpar, botrles
2. Add accompanying vegetables, a drink and a dessert to these fish to make them into well-balanced meals.
   a) stuffed baked fish
   b) sardines on toast
   c) fish and chips
   d) pilchard salad
   e) grilled fish fingers
   f) cod in cheese sauce
   g) fried fish cakes
   h) grilled kippers
3. Explain where and for how long the following foods should be stored:
   a) fresh plaice fillets
   b) frozen plaice fillets in a refrigerator star marking **
   c) fish fingers in the deep freeze
   d) frozen cod steaks in a fridge marked ***
4. Fish and chips are a popular traditional British food. These are the names of some other fish dishes. Find out what they are and where they come from:
   Finnan haddock, Arbroath smokies, kedgeree, stargazy pie, rollmop herring

Stuffed baked fish

Ingredients
(*for 4 people*)
4 steaks of white fish (cod, hake, coley)
50 g white or brown bread
½ lemon
50 g butter or margarine
fresh parsley or 1 tsp mixed herbs
salt, pepper
lemon and parsley to garnish

Utensils
scissors, vegetable knife, cook's knife, board, teaspoon, wooden spoon, grater, plate, basin, saucepan, ovenproof dish

Method
Collect ingredients and utensils. Light oven at gas no. 5 (190°C, 375°F). Grease an ovenproof dish.

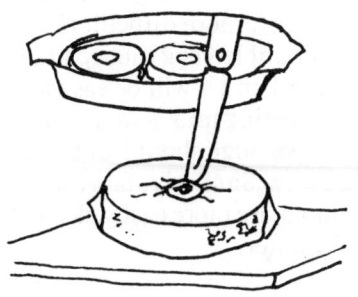

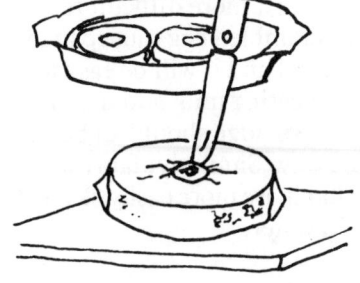

1 Wash the fish, dry on kitchen paper. Trim any fins and remove the centre bone using scissors or a vegetable knife. Lay the steaks side by side in the ovenproof dish.

2 Prepare the stuffing: Remove the crusts from the bread. Make into breadcrumbs by rubbing on the medium-sized holes of the grater on to a plate.

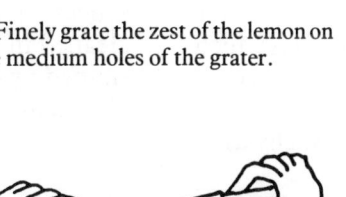

3 Finely grate the zest of the lemon on the medium holes of the grater.

4 Chop the parsley, use enough to make 1 tblsp.

5 Mix the crumbs, lemon and parsley (or mixed herbs) together in a basin. Add the seasoning.

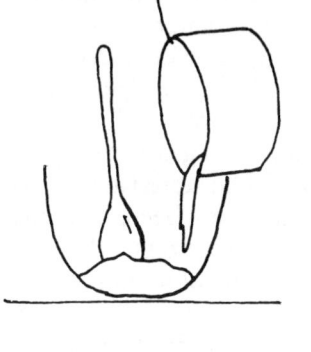

6 Melt the fat in a saucepan. Pour it on to the crumb mixture and mix to a stiff paste.

7 Divide stuffing into four. Use a teaspoon to neatly fill each hole in the fish steaks.

8 Bake for 20 minutes, until the fish is tender.

Serve hot with boiled or mashed potato and a colourful vegetable, e.g. baked tomato.
  Use the oven to bake a pudding at the same time, e.g. baked apples.

Eggs

The eggs most commonly used in this country are hen's eggs. They can be used in a number of ways, on their own or in a variety of different dishes.

The egg provides all the food needed for the growth and life of the chick until it is hatched from its shell. This food is also very useful for humans as it contains protein for growth and repair, fat for energy and micronutrients for protection and maintenance (Vitamins A, D, B, iron, calcium and phosphorus). The egg also contains water.

Eggs may be stored for up to six weeks in a cool place, a refrigerator is not needed. The egg shell is porous, which means that it absorbs smells and gases from its surroundings. Eggs should be stored away from strong smelling foods, e.g. cheeses, onions.

Different types of hens lay eggs with different coloured shells but the food inside the egg is the same; a white egg is just as useful as a brown one. Hens lay eggs of varying sizes which are graded into six groups according to their weight. Size 1 is the largest, heaviest egg, size 6 is the smallest. Size 3 is the egg size usually used in most recipes.

What happens when an egg is cooked?

An uncooked egg is a thick, runny liquid. When heated it changes and becomes thicker, then sets and becomes firm. If overheated it becomes hard and rubbery. All these changes are due to the effect of heat on protein. When the protein in a food is heated, it changes and sets. This setting of protein is called **coagulation**. This is an important change in the cooking of all foods containing protein. If overcooked, the protein becomes too hard: meat and fish become tough and stringy, a skin forms on milk and cheese becomes stringy.

Utensils that have been used to cook protein foods are more easily cleaned if they are soaked in cold water. Hot water would only 'cook' the protein further, making it more difficult to remove. Leave the scrambled egg saucepan in cold water while you eat and it will be easy to clean. Pans used for heating milk and making sauces, custards and porridge should all be cleaned in the same way. Similarly, fabrics stained by protein foods are more easily cleaned after soaking in cold water.

Things to do

1 Crack an egg into a saucer. Look at these parts:
the shell, the membrane lining the shell, the air sac, the thin white, the thick white, the egg yolk, the egg strands
Draw and label the diagram of the egg:

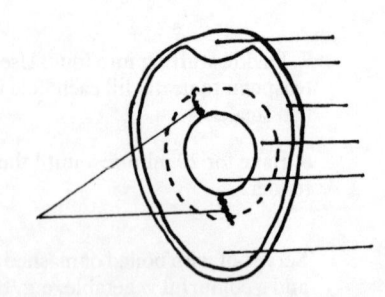

2 Make a list of the ingredients needed to prepare scrambled egg for a) two people b) four people.

3 Complete these sentences:
a) Eggs are a useful food because they contain p------, f--, v------s, i---, c------, p-------u-.
b) There are many ways of cooking eggs, e.g. bo-----, f----g, s-r-------, p-------.
c) The protein in egg c----l---- when heated.
d) Eggs may be stored for ___.

4 These are some dishes made with eggs. Make them into well-balanced meals by adding vegetables, drinks and desserts where needed.
a) stuffed eggs b) poached eggs on toast c) cheese omelette d) scrambled egg e) Scotch egg f) baked eggs in cheese sauce

> *New word*
> **coagulation** – the setting or hardening of foods containing protein as a result of cooking or heating

Scrambled egg

Ingredients
(for 1 person)
1 egg
2 tblsp milk
25 g butter or margarine
pinch of salt, pepper
1 slice of bread
parsley

Utensils
basin, saucepan, wooden spoon, fork, teaspoon, knife, serving plate

Method
Collect ingredients and utensils. Make the toast. Put toast on serving plate to keep warm.

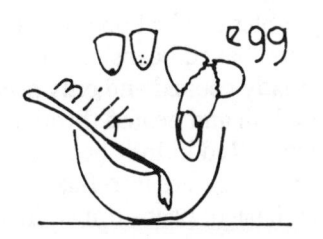

1 Break egg into a basin. Add seasoning and 1 tblsp milk.

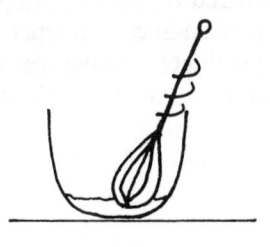

2 Mix lightly with a fork or wire whisk.

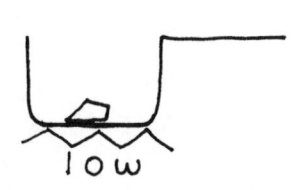

3 Melt half the butter or margarine in a saucepan on a low heat.

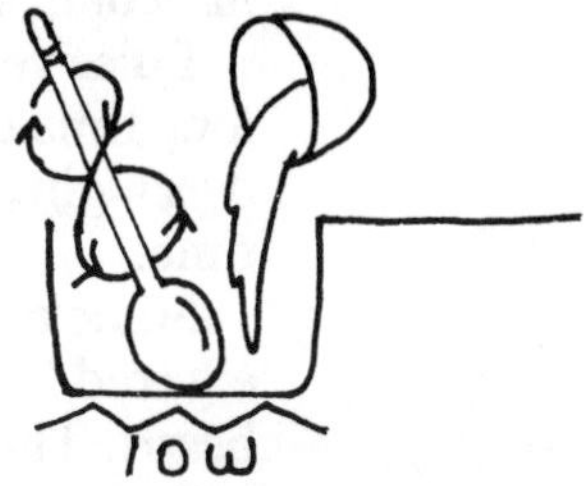

4 Add the egg mixture. Stir in a figure of 8, moving the setting egg from the base of the pan as it cooks.

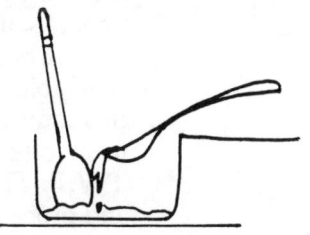

5 As soon as it is thick, add the second tblsp of milk. Stir well and remove from heat at once. The heat of the pan will continue to cook the egg. Scrambled egg should be smooth and creamy, not lumpy and rubbery.

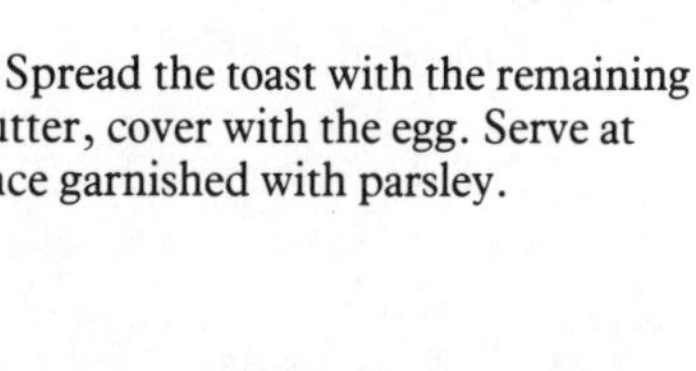

6 Spread the toast with the remaining butter, cover with the egg. Serve at once garnished with parsley.

Scrambled egg becomes rubbery if left before serving. The table should be laid and the rest of the meal ready to serve before the egg is cooked, so that it is served hot and freshly cooked.

Scrambled egg on toast makes a versatile snack or meal at any time of the day. (Suggested menus: breakfast – grapefruit, bacon, scrambled egg, coffee; lunch – scrambled egg on toast, milk shake, apple or orange; supper – soup, cheese scrambled egg on toast, flapjack)

Pulse vegetables

Pulses are the seeds of certain plants. Beans, peas, kidney beans, lentils, haricot beans and soya beans are all pulses. They contain starch, some micronutrients and sometimes fat, but unlike most other vegetables they are a good source of protein.

Protein is needed by the body for growth and repair. It is obtained in the diet from animal flesh and products, which contain protein.

Some vegetables also contain protein.

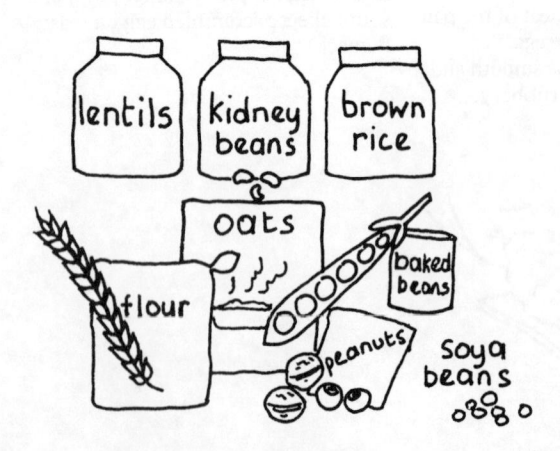

The protein from vegetable sources is not as useful to the body, when eaten alone, as that from animals. However, when the two types are eaten together they are very useful, e.g. cornflakes (vegetable protein) and milk (animal protein), ham and pease pudding.

Since animal protein foods are often more expensive, pulses and vegetable protein can make them go further, e.g. lamb stew with haricot beans.

Pulse vegetables are usually hard and take a long time to cook. They may need to be soaked overnight; the use of a pressure cooker saves time.

Some pulse vegetables can be bought tinned, e.g. red kidney beans, baked beans, butter beans. They are already cooked and only need to be brought to the boil and heated through thoroughly before use. Red kidney beans must be boiled for 10 minutes before use. Baked beans are processed haricot beans tinned in a tomato sauce.

Soya beans are the most useful pulse vegetable, though they are not eaten alone in this country. They can be used to make artificial meat, often called textured vegetable protein or T.V.P. This is widely used in convenience and processed meat products to make real meat go further. The label must clearly state the meat content.

Special baked beans

Most dishes using pulse vegetables take a long time to cook. This is a quick dish using tinned beans that are already cooked.

Ingredients
(for 2 people)
1 small tin baked beans
½ onion
2–3 rashers streaky bacon
25 g margarine
salt, pepper
½ tsp brown sugar
½ tsp vinegar
Worcester sauce
2 slices bread

Utensils
scissors, vegetable knife, board, saucepan, tin opener, teaspoon, serving plates

Method
Collect ingredients and utensils. Make toast. Leave plates and toast to keep warm.

1 Remove rind from bacon with scissors. Chop finely.

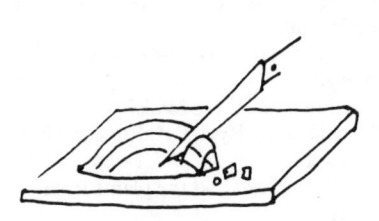

2 Peel and finely chop the onion.

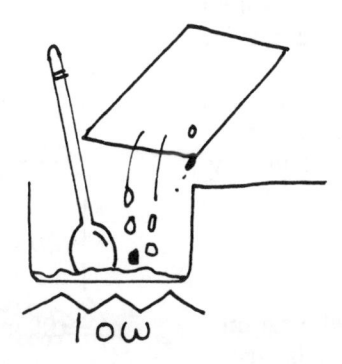

3 Melt half the fat in a saucepan. Gently fry bacon and onion for 5 minutes.

4 Open the tin of beans, add to the bacon/onion mixture. Add seasoning, sugar and vinegar, and a few drops of Worcester sauce. Heat through thoroughly.

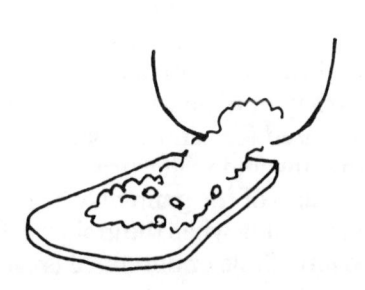

5 Spread remaining margarine on the toast and cover with beans. Serve at once.

Cereals

Cereals are the seeds of a group of grass-like plants which include wheat, oats, barley, rye, rice, maize. Cereals have always played an important part in the diet, so much so that the Ancient Greeks worshipped a goddess of corn called Ceres, from whose name the word cereal is taken.

Cereal plants contain large amounts of starch. This is used as a food store by the plant, and when digested, provides the human body with energy. Cereal products are usually filling and form the basis of the diet. In Britain, wheat is grown and used for flour; in China and Japan, rice is the staple cereal; in Russia rye is used for flour; and in Italy a special type of wheat is used to make spaghetti.

In addition to starch, cereals contain protein and small amounts of Vitamin B, calcium, iron, Vitamin E and dietary fibre.

The different cereals are processed or ground in different ways to make them suitable for a variety of uses: wheat is ground into flour, oats are used for oatmeal, barley is used for beer making and malt extract.

Parts of the grain may be removed during processing when the grain is refined. Grain that is left whole or unrefined is called wholegrain and retains the natural nutrients and dietary fibre. It is useful in the diet for this reason.

The word cereal is often linked with food eaten at breakfast, e.g. 'cornflakes' which are made from maize (also sometimes called corn). The manufacturing process removes the dietary fibre. Breakfast cereals provide starch for energy, and when taken with milk to provide protein and fruit to provide vitamins, they can be a filling and balanced breakfast to prepare and eat quickly.

Only a little sugar should be added, especially if the cereal is already sweetened. Breakfast cereals containing dietary fibre should also be eaten, e.g. porridge, muesli, bran products.

Things to do

1 Draw the diagram below. In each balloon draw a food which is made from the cereal in the box attached to it.

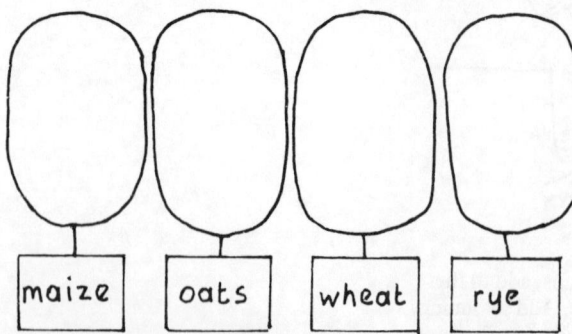

[maize] [oats] [wheat] [rye]

2 Choose the correct word from the words in brackets to complete the sentences.

a) Cereals are the grains of —— (*animals, plants, minerals*)
b) The staple cereal in Britain is ——. (*rice, oats, wheat*)
c) When refined, the —— is removed from the cereal grain. (*starch, fibre, fat*)
d) Cereals are useful in the diet because they give ——. (*energy, fat, strength*)

3 Which cereal is used to make these foods?
a) spaghetti b) 'Weetabix' c) 'Ryvita' d) Russian black bread e) pumpernickel

4 Make a list of breakfast cereals and the plants from which they are made.

5 What is a) corn on the cob b) sweetcorn c) popcorn? Find out how they can be bought, cooked and served.

6 Find the recipes for these dishes made with cereals: flapjacks, porridge, muesli, bread, wholemeal pastry, lemon and barley water, barley soup

Muesli

Ingredients
(for 2 people)
2 level tblsp rolled oats
juice of ½ orange (2 tblsp)
1 red-skinned eating apple
2 tblsp milk
1 dsp honey
25 g sultanas or raisins
25 g chopped nuts

Utensils
lemon squeezer, grater, plate, basin,
teaspoon, tablespoon, serving dishes

Method
Collect ingredients and utensils.

The muesli can be eaten straight away, or can be left overnight in the refrigerator. Serve for breakfast or as a dessert at the end of a meal.

1 Place oats in a basin. Add orange juice, stir to mix. Leave to soak for 20 minutes.

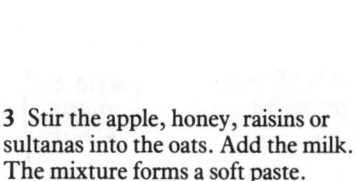

3 Stir the apple, honey, raisins or sultanas into the oats. Add the milk. The mixture forms a soft paste.

2 Meanwhile, wash the apple. Grate it on to the plate on the large holes of the grater.

4 Spoon into serving dishes. Sprinkle with the chopped nuts.

Flour

Flour is ground from a plant called wheat. Grinding takes place in a mill where the grains may be crushed by steel rollers or more old-fashioned stones, producing 'stone-ground' flour. Windmills and water-mills have both been used for this purpose.

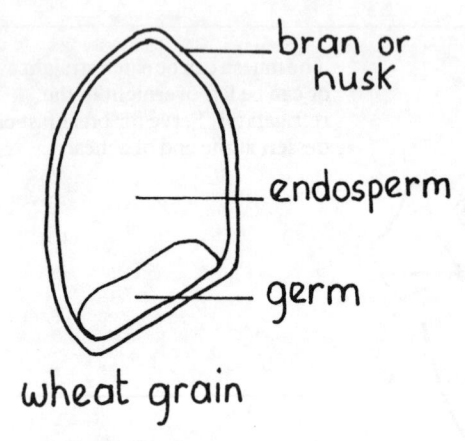

leaving the kernel only. Calcium, iron and Vitamin B are added to white flour to make up for the losses, but it does not contain dietary fibre. Useful for some cakes, pastries, sauces.

Self raising flour
This has chemicals added to it to produce the gas carbon dioxide when a mixture is baked. This makes the mixture rise, producing a light result.

Wholemeal flour
The whole grain is ground into flour. It is brown in colour as it contains the bran and germ. The bran contains dietary fibre which is important for a healthy digestive system. It can be used in many recipes instead of white flour, e.g. pastries, cakes, bread.

White flour
The wheat grain is refined. This means that the bran and germ are removed during milling

Things to do

1. Draw and label a diagram of a grain of wheat.
2. Design a poster to advertise wholemeal bread.
3. Copy these sentences, explain whether each one is *true* or *false*.
   a) Wholemeal flour is white in colour.
   b) Wholemeal flour can be used for making cakes.
   c) White flour contains dietary fibre.
   d) White flour contains calcium, iron, Vitamin B.
   e) Wholemeal flour contains calcium, iron Vitamin B.
   f) White flour contains wheat germ.
   g) The chemical that makes soda bread light is bicarbonate of soda.
4. If a recipe tells you to knead a mixture, what will you do?

Soda bread

Ingredients
200 g plain white or wholemeal flour
1 level tsp salt
1 level tsp bicarbonate of soda
1 level tsp sugar
15 ml sour milk or a 5 oz carton plain yoghurt and 1 tblsp milk

Utensils
mixing bowl, sieve, tablespoon, teaspoon, fork, measuring jug, flour sifter, baking tray

Method
Light the oven at gas. no. 7 (elec. 210°).

1 Sieve the flour, salt and bicarbonate of soda into the bowl. Add the sugar.

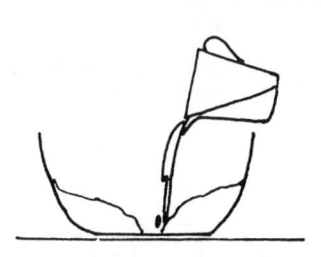

2 Make a well or dip in the flour. Add the milk.

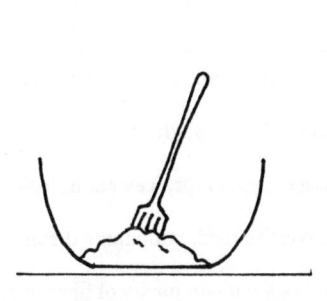

3 Use a fork to mix to a stiff dough.

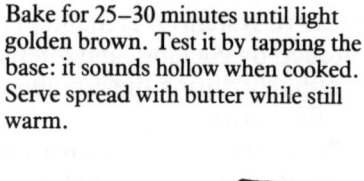

4 Place dough on a table lightly sprinkled with flour to prevent it sticking. **Knead** lightly until smooth.

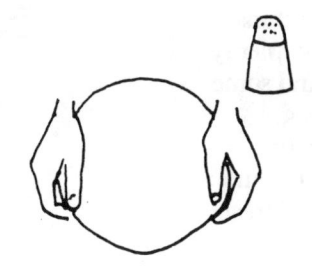

5 Shape into a circle with a smooth surface.

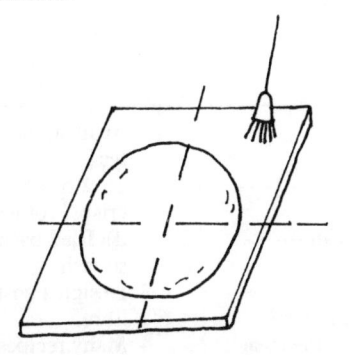

6 Place on a greased baking tray. Mark into quarters with a sharp knife.

Bake for 25–30 minutes until light golden brown. Test it by tapping the base: it sounds hollow when cooked. Serve spread with butter while still warm.

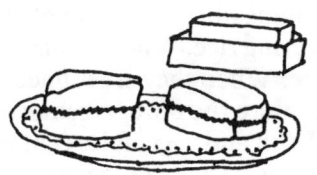

New word
knead – to work a mixture with the hands, either in a bowl or on the table (pastry is kneaded lightly with the fingertips, bread is kneaded firmly using the whole hand, as with plasticine or clay in pottery)

Dietary fibre

When food is eaten, much of it is digested to give materials for growth, maintenance, repair and protection, and to provide energy.

The human digestive system cannot digest a substance called cellulose, it passes through the system undigested. Cellulose, or dietary fibre, gives bulk to the food and helps waste to pass through the system, preventing constipation. It is important for the general health of the digestive system. It is found in all plant food, is softened by cooking, and often removed completely during the refining of some foods, e.g. flour, rice.

The outer covering of many cereal grains contains fibre, e.g. bran of wheat, the husk covering rice. Whole grains are a good source of fibre. When they are refined, the covering, and therefore the fibre, is removed.

Fruit and vegetables contain plenty of dietary fibre, especially in the skins, though of course, this part cannot always be used. However, new potatoes boiled in their jackets and baked potatoes eaten with their skins are good sources of fibre.

Pulses and nuts contain fibre in their skins.

To keep the digestive system healthy, plenty of raw or cooked fruit and vegetables and some unrefined food should be eaten daily, e.g. wholemeal bread; use wholemeal flour for pastry and cakes; use brown rice; eat nuts and dried fruit in place of sweets; eat a piece of fresh fruit in place of biscuits and cakes.

Things to do

1 Complete these sentences:
  a) Dietary fibre is found in all –l––––.
  b) It is needed in the diet for a healthy d–––––––– s–––––.
  c) Wholemeal flour contains more fibre than –––t–f––––.
  d) The fibre in wheat is found in the –r––.
2 Which foods in these meals contain dietary fibre? Underline the foods containing fibre. Alter the meal if necessary to include fibre.
  a) stuffed egg, mixed salad, peach melba, fruit juice
  b) meat pie, baked jacket potato, cabbage, fruit jelly, tea
  c) White bread ham sandwich, yoghurt, chocolate bar, crisps, coffee
  d) fried bacon, sausage, egg, chips, ice-cream, fruit squash
3 Design a poster to advertise foods containing dietary fibre.
4 Many recipes in this book contain plenty of fibre, e.g. flapjack, salads, muesli. Make a list of all the recipes you can find that contribute fibre to the diet.

Wholemeal scones

Ingredients
150 g wholemeal flour
50 g plain white flour
pinch of salt
1 level tsp baking powder
50 g margarine
150 ml milk
50 g caster sugar
50 g sultanas

Utensils
mixing bowl, sieve, palette knife, teaspoon, flour sifter, rolling pin, baking tray, cutter, brush

Method
Collect ingredients and utensils. Grease a flat baking tray. Light the oven at gas. no. 7 (220°C, 425°F).

1 Sieve the white flour, salt and baking powder into mixing bowl.

2 Add the wholemeal flour. Do not sieve this (the bran will not pass through the sieve).

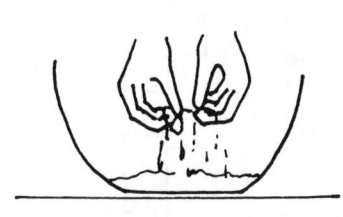

3 Cut the fat into small pieces. Crumble it into the flour by rubbing-in between the finger tips until the mixture looks like fine crumbs.

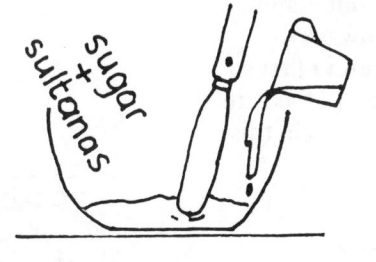

4 Add the sugar and sultanas. Stir in the milk, mixing with a palette knife. Knead with the fingertips to form a soft dough.

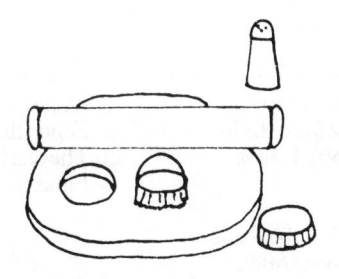

5 Lightly flour the table and rolling pin. Roll out the dough about 1 cm thick. Do not roll it too thinly. Cut into circles. Glaze the tops with milk.

6 Bake for 15–20 minutes until risen and golden brown.

Serve warm or cold with butter. Cheese scones can be made by using 50 g grated cheese instead of sugar and sultanas.

Starch

Starch is found in many foods, e.g. flour, cereals, potatoes, spaghetti. Foods that contain starch are often filling and provide energy and warmth. Starch has many uses in cooking: it is often used to make liquids thicker, to make custard and sauces, or to set liquids firmly, e.g. blancmange. The starches most often used for thickening are: flour, cornflour, blancmange powder, arrowroot, custard powder.

If starch is mixed with a liquid and heated it becomes thick as the starch grains burst and absorb the liquid. It is important always to blend the starch with a little cold liquid first so that it thickens slowly. Some packets of food are labelled to show they contain edible starch, this is pure starch.

Why milk boils over

Milk contains protein. When heated, protein becomes hard and sets.

1 When milk is heated, milk protein hardens to form a skin on the surface.
2 The water in the milk begins to evaporate and turn into steam.
3 More steam is produced but cannot escape as it is trapped by the skin.
4 Steam builds up pressure under the skin, forcing it up as the milk boils, and causing the milk to boil over. Watch the milk heating to prevent this happening.

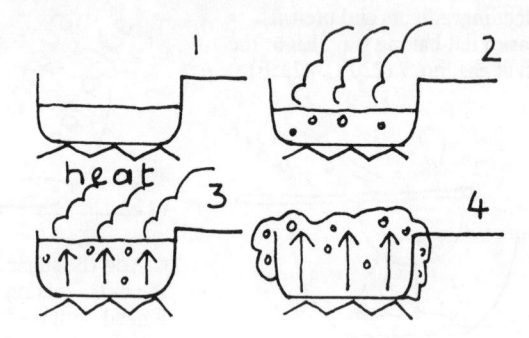

Things to do

1 Try this experiment. Copy the instructions into your book. Fill in the missing words, choosing from the list.
 a) Place 2 tblsp flour in a basin. Add 2 tblsp boiling water. What do you see?
 The mixture becomes ——.
 b) Place 2 tblsp flour in another basin. Add 2 tblsp cold water. Blend smoothly with a wooden spoon. Add 2 more tblsp cold water and heat gently.
 The mixture becomes ——.
 thin, lumpy, glue-like, thick, smooth, sticky
2 Look at the packets containing these foods and make a list of the contents of each:
    custard powder, gravy powder, blancmange
3 Look up the word edible in the dictionary. What does it mean? Make a list of foods that contain edible starch.
4 Which starch would you use to thicken the following?
 a) cheese sauce b) porridge c) glaze for fruit flan
 d) lemon meringue pie
5 Fill in the crossword.

Clues across
1 Starch and liquid become like this when heated.
3 Barley is used to make this drink.
4 Found in all cereal plants.
5 Used to make bread.
6 Plant used to make 5 across.

Clues down
2 Foods that give us energy and warmth.
3 The staff of life.
4 Flour is used to thicken these.

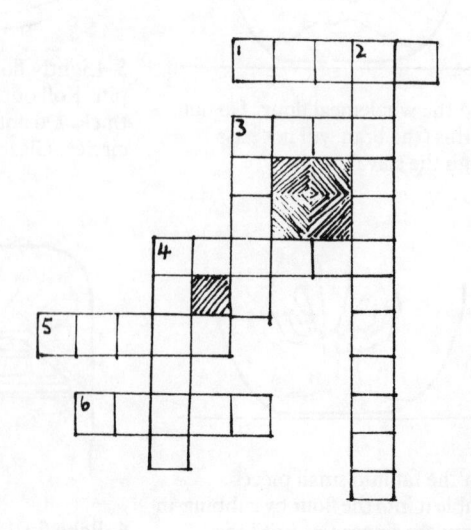

50

Fruit fool

Ingredients
600 ml milk
2 rounded tblsp cornflour
2 tblsp sugar
300 ml fruit purée made with any soft fruit, e.g. plums, strawberries, apples
2 tblsp double cream

Utensils
measuring jug, sieve or liquidiser, wooden spoon, saucepan, basin, tablespoon, serving dish

Method

1 Measure 600 ml milk in a jug or bottle.

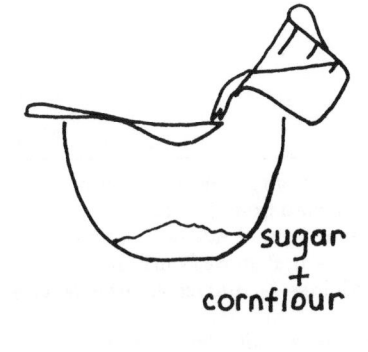

2 Place cornflour and sugar in basin. Add 2 tblsp cold milk from the measured 600 ml. Blend together with a wooden spoon so that there are no lumps.

3 Heat the remaining milk in a saucepan; do not let it boil. When it is steaming, pour the milk over the cornflour mixture. It will begin to thicken as the starch begins to absorb some of the hot milk. Pour back into the saucepan and bring to the boil to finish thickening.

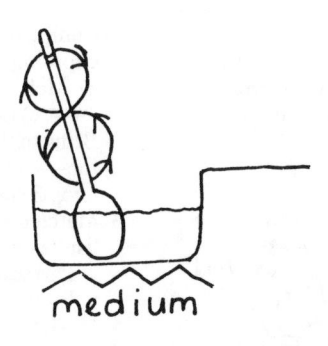

4 Stir in a figure of 8 until thick and smooth.

5 Leave the mixture to cool quickly in a bowl of cold water. To prevent the milk from forming a 'skin' as it sets, cover with wet greaseproof paper.

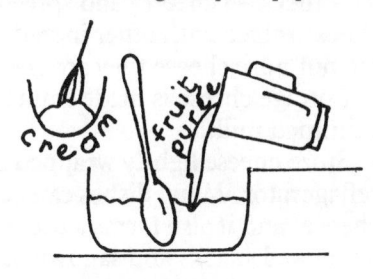

6 Meanwhile, prepare the fruit purée. Stir 300 ml fruit purée into the cool mixture. Beat the cream until thick and stir into the fool.

Spoon the fool into serving dishes and chill. Serve with a crisp biscuit.
    Fruit fool should be soft, not thick and rubbery.

Cheese

Cheese is a useful food made from milk. Cow's milk is generally used in this country, though goat's milk and ewe's milk cheeses are eaten in other countries.

Milk is a liquid which contains mainly water. To prepare cheese, the water is removed leaving the nutrients in a concentrated form. These nutrients are protein, calcium, phosphorus, fat, Vitamins A and D. During cheese manufacture the milk is separated into two parts: the liquid whey and the solid curd. The curd is pressed to make cheese.

The main types of cheeses are:
a) Hard cheeses – Cheddar, Cheshire, Parmesan, Edam, Gloucester.
b) Soft cheeses – Caerphilly, Camembert, Brie. Some contain coloured moulds giving a strong smell and flavour – Gorgonzola from Italy, Stilton from England, Danish Blue.
c) Cream cheeses – made from cream. These are rich, soft and easy to spread.
d) Processed cheeses and spreads – made from cheese mixed with other ingredients. As they are not pure cheese they are cheaper.

Cottage cheese is made from fat-free or skimmed milk.

Store cheese lightly wrapped in the refrigerator. Many dishes can be made using cheese, and it also forms a useful basis for uncooked snacks and sandwiches.

Sauces

Sauces add flavour and variety to many foods.

A sauce is a thickened liquid, often made with stock or milk. It may be served with a meal (e.g. parsley sauce with fish) or as part of a dish (e.g. macaroni cheese). The liquid may be thickened with cornflour, or flour mixed with an equal quantity of fat to form a paste called a **roux**. The sauce varies in thickness depending on the proportion of flour to liquid: 15 g flour to 250 ml liquid will make a thin or pouring sauce; 25 g flour to 250 ml liquid will make a thick or coating sauce.

Follow these points carefully to prevent lumps from forming when making a sauce thickened with a roux:
a) Melt the fat, remove from heat, add the flour. Mix to a paste — this is the roux.
b) Cook the roux gently for two minutes.
c) Remove from heat, add half the liquid, mix well.
d) Heat gently, stir in a figure of 8 until the sauce begins to thicken. Add remaining liquid and continue cooking for one minute.

Add seasoning and flavouring when the sauce is cooked. When using cheese, do not boil the sauce again or the cheese will become stringy.

Things to do

1 Experiment to make cheese.
   You will need 300 ml milk, 1 tblsp vinegar, muslin, a sieve. Heat the milk and vinegar gently in a saucepan. It will begin to separate and curdle. Place the muslin in the sieve over a jug. Pour in the milk.

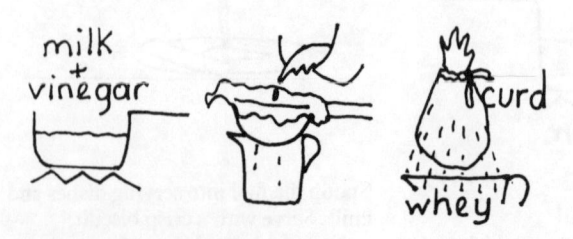

Tie the muslin to make a bag, leave to drain the curds.

2 Complete these instructions for making a sauce, choosing the correct word from the list.
   a) Melt —— in a saucepan. b) Add ——. Stir to form a ——. c) Cook for ——. d) Remove from heat, add ——. Bring to the boil, stir in ——. Add remaining milk, cook for ——.
   roux, 2 minutes, one minute, milk, fat, figure of 8, flour
3 Match the cheeses with their country of origin from the list below:
   Parmesan, Edam, Brie, Stilton, Gruyère, Gorgonzola
   England, France, Italy, Holland, Switzerland
4 Mark the areas of origin of these cheeses on a map of Britain:
   Leicester, Stilton, Cheddar, Cheshire, Caerphilly, Double Gloucester

Macaroni cheese

Ingredients
(for 2–3 people)
100 g macaroni
250 ml milk ⎫
25 g margarine ⎬ coating sauce
25 g flour ⎭
¼ tsp each salt, pepper
100 g grated cheese
25 g breadcrumbs (1 slice bread)
parsley or tomato

Utensils
saucepan, fork, wooden spoon, teaspoon, colander, 500 ml ovenproof serving dish

Method
Collect ingredients and utensils. Grease an ovenproof dish.

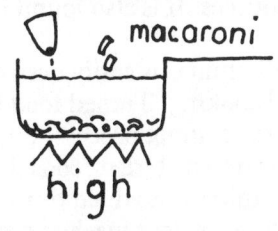

1 Place macaroni in saucepan, cover with plenty of cold water, add 1 tsp salt. Bring to the boil on high heat.

2 Lower the heat and boil gently for 20 minutes, or until soft. Stir occasionally with a fork to prevent it sticking.

3 Meanwhile, make the cheese sauce. Melt margarine in a saucepan over a low heat. Remove from heat and stir in the flour to make a roux.

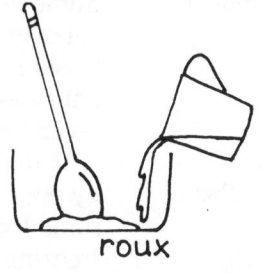

4 Cook the roux gently for 2 minutes. Remove from the heat and add half the milk, stirring smoothly.

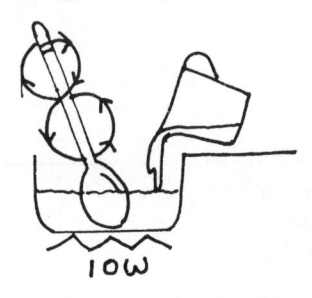

5 Heat the sauce gently, stirring all the time in a figure of 8 until it becomes thick and smooth. Add remaining milk, boil for one minute.

6 Remove from heat, add seasoning and ¾ of the grated cheese. Stir in the well-drained macaroni.

7 Pour into the serving dish. Mix breadcrumbs and remaining cheese and sprinkle over the surface. Brown under the grill.

Serve hot, garnished with tomato or parsley for lunch or supper. Add toast or a crisp vegetable to give a change of texture. (Suggested menu: macaroni cheese, toast triangles, tomato, apple.)

Cheese sauce may be used to coat other foods, e.g. fish, hard boiled eggs, boiled cauliflower. The cheese coated dish is sometimes called 'mornay', e.g. fish mornay.

Other sauces may be made by adding different flavourings, e.g. parsley sauce (½ tblsp chopped parsley), egg sauce (½ chopped hard boiled egg), mushroom sauce (25 g chopped cooked mushrooms), onion sauce (1 boiled onion).

New word
roux – equal mixture of flour and fat used to thicken sauces

Fruit and vegetables

Fruit and vegetables add variety of colour, flavour and texture to the diet. They can be eaten raw or cooked, and contain **micronutrients** needed by the body for repair, maintenance and protection. These micronutrients are the vitamins and minerals needed in small but vital amounts by the body for various functions:

a) Vitamin A for health and good eyesight
b) Vitamin B for a healthy nervous system
c) Vitamin C (ascorbic acid) for prevention of scurvy and general health
d) Vitamin D for bones and teeth
e) Iron for healthy blood
f) Calcium for healthy bones and teeth

Apart from iron and calcium, iodine, phosphorus and fluorine are among the other minerals needed in the diet.

These micronutrients are found in many other foods as well as fruit and vegetables, e.g. calcium in milk and cheese, iron in meat and eggs.

Fruit and vegetables contain fibre which is softened by cooking. Raw fruit and vegetables are a good source of fibre in the diet; a piece of raw fruit should be eaten daily. The crisp texture of most of them is good exercise for gums and teeth, so fruit is a better snack than sweets or sticky buns and cakes.

Different fruits contain different micronutrients, but in general they are a useful source of Vitamin C, especially when eaten raw. Some food containing Vitamin C should be eaten every day – blackcurrants, rosehips (rosehip syrup), strawberries and the citrus fruits are all good sources. It is also found in green vegetables.

Unfortunately, Vitamin C is easily destroyed by poor storage and cooking. Tinned food loses some Vitamin C in the tinning process, but freezing does not destroy it. Dried foods do not contain Vitamin C, unless it is added during manufacture, e.g. some instant mashed potato.

Things to do

1 Design a poster to sell fresh fruit and vegetables.
2 Which of these foods contains most Vitamin C? Rearrange each list in order, starting with the food containing the most:
   a) chips, tinned new potato, boiled new potato, boiled old potato
   b) celery, tomato, orange, meat, sugar
   c) boiled cabbage, raw cabbage, steamed cabbage
3 Suggest vegetables to accompany these dishes:
   a) sausage hot pot b) beef casserole c) mixed grill
   d) roast beef e) cheese flan
4 Fill in the crossword.

Clues down
1 Fruit and vegetables must be bought like this.
2 (and 8 across) Chemical name for Vitamin C.
3 Needed for healthy teeth and bones.
6 Fruit rich in Vitamin C.

Clues across
4 Group of fruits rich in Vitamin C.
5 Scent of food cooking.
7 Do this to extract flavour.
8 See 2 down.

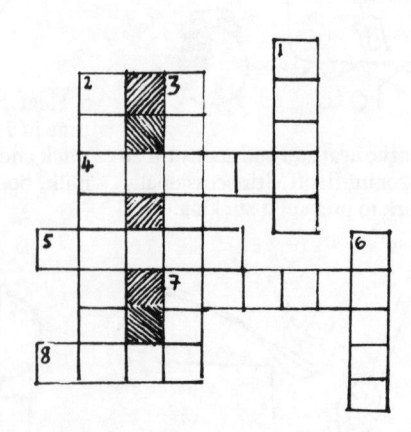

New word
micronutrients – food substances needed in small but vital amounts for healthy functioning of the body

Stuffed tomatoes

Ingredients
(for 4 people)
4 large firm tomatoes
50 g cream cheese
1 tblsp milk
2 spring onions or a bunch of fresh chives
lettuce or watercress to garnish

Utensils
teaspoon, mixing bowl, plate, wooden spoon, vegetable knife, serving plate

3 Beat cheese, milk, seasoning and chopped onions or chives in a basin. This makes a thick creamy filling.

Method
Collect ingredients and utensils.

1 Wash the tomatoes. Cut a slice off the top of each one, keep the top for decoration. Use a teaspoon to scoop out the insides.

2 Sprinkle inside the tomato cases with salt, leave upside-down to drain on a plate for 20 minutes. The salt drains water from the tomato.

4 Divide the filling into 4. Use a teaspoon to fill each tomato case with the mixture. Replace 'caps' on each tomato. Chill before serving on lettuce leaves or surrounded with watercress.

Eat with brown bread and butter and a salad.

Coleslaw

Ingredients
(for 4 people)
½ white cabbage
2 carrots
2–3 tblsp mayonnaise
25 g raisins ⎱ optional
1 eating apple ⎰

Utensils
chopping board, cook's knife, grater, basin, tablespoon, salad or serving bowl

Method
Collect ingredients and utensils.

1 Cut the cabbage into 2. Remove the centre stem. Shred finely using a cook's knife on a chopping board.

2 Wash, peel and coarsely grate the carrots on the large holes of a grater.

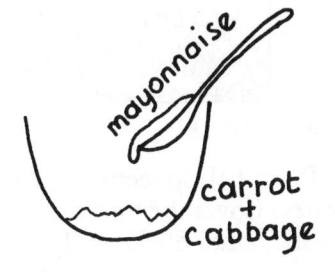

3 Mix cabbage, carrots and mayonnaise to bind together. Add the raisins and chopped apple if used. Coat apple with lemon juice.

Serve as soon as possible in dishes or salad bowls.

Grow your own salad

The seeds of some plants sprout quickly to give shoots that can be used to add crispness, variety, and Vitamin C to mixed salads.

The most commonly used seeds are mustard and cress, which can also be used as a garnish for savouries and sandwiches. Some more unusual seeds for sprouting include mung beans and alfalfa. Mung beans produce bean sprouts which can be used raw in salads.

Mung beans
These will take about four days to sprout.
You will need:
1 packet mung beans
1 large glass jar
muslin or woven paper kitchen cloth

Method

1 Place 1 tblsp seeds in the jar. Fill with warm water. Cover the top with the cloth tied with an elastic band. Shake well.

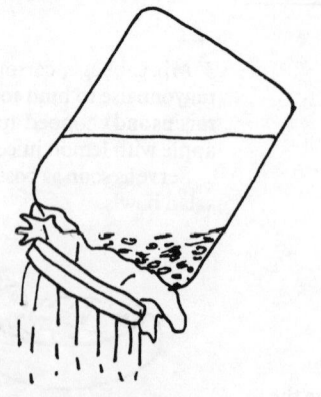

2 Pour off the water through the cloth.

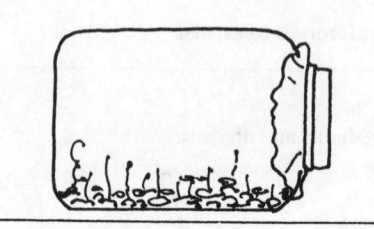

3 Leave the jar on its side, rinse the seeds with warm water daily in the same way until they fill the jar. Use in salads, store any extra in a covered box in the refrigerator.

Mustard and cress
These will take a few days to grow.
You will need:
1 packet mustard and cress seeds, plate, blotting paper or new flannel.

Method

Wet the paper or flannel. Lay it on the plate sprinkled with the seeds. Leave in a cool light place (e.g. windowsill), keep moist and the seeds will sprout fast. Cut off the tops and use on salads.

Mixed salad

Ingredients
(*for 2–3 people*)
1 small lettuce
¼ cucumber
2 tomatoes
1 bunch radishes
other vegetables may be used:
  watercress, mustard and cress,
  peppers, bean sprouts, parsley

Dressing
2 tblsp vegetable oil
1 tblsp vinegar
¼ tsp each of salt, pepper, mustard powder

Utensils
colander, clean tea-towel, vegetable knife, board, basin and fork or jar, serving bowl or plate

Method
Collect ingredients and utensils.

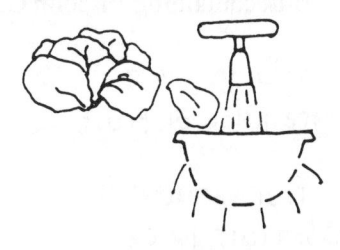

1 Remove outer or damaged leaves of lettuce. Tear the leaves to separate them, place in a colander. Rinse under cold water to remove soil and dirt.

2 Shake off surplus water. Dry the leaves thoroughly on a clean tea-towel.

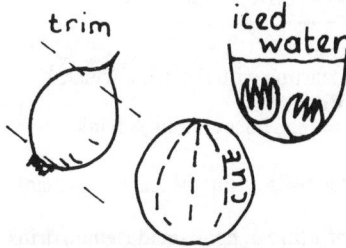

3 Trim the radishes. Wash and scrub if necessary. Leave whole if small, or cut into slices, or make a radish 'rose' (cut into the radish as shown, leaving the end uncut, leave in iced water for ½ hour until it has opened).

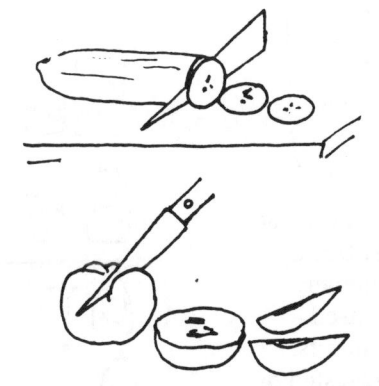

4 Wash and slice or dice the cucumber. Wash and cut the tomatoes into quarters.

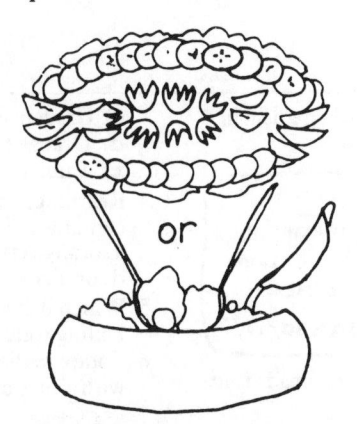

5 Place the prepared vegetables in a salad bowl or arrange neatly on a plate. Cover and chill, use as soon as possible coated with dressing.

Dressing

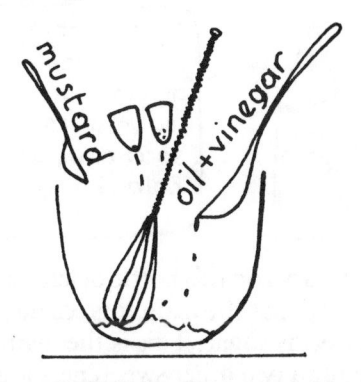

Pour the oil and vinegar into a basin. Add the seasoning, whisk with a fork or wire whisk until mixed.

or
Place dressing ingredients in a jar with a screw top, shake together to mix.

Pour the dressing over the salad just before serving. Toss the salad in the bowl until it is all coated. Serve the salad as an accompaniment to a hot dish, e.g. spaghetti bolognese, or add protein (eggs, cheese, cold meat or fish) to make a cold meal, e.g. sardine and hard boiled egg salad.

Fruit drinks

When food and drinks are prepared at home, you know what they contain. But how do you know what prepared or convenience foods contain? Some orange drinks need not contain real oranges at all.

Read the label on the food or drink. All food and drink must be clearly labelled with the ingredients, except water. The label need not state the amounts of ingredients, but they must be listed in order of weight, starting with the heaviest.

Ingredients: sucrose, lemon, lime, glucose syrup, fruit acids, sodium citrate, stabiliser, saccharin, preservative, colour

The label on this bottle of lemon and lime drink shows that the main ingredient, apart from water, is sugar. This is the main sweetener, though two other sweeteners are also used – glucose and saccharin – in smaller amounts. The main fruit flavour is lemon, as suggested by its name where the word lemon is used first. This is also shown in the list of ingredients.

The labels on fruit drinks will tell you whether they contain real fruit, and if so how much.

Some fruit drinks, squashes etc. may contain Vitamin C. This is sometimes called by its chemical name, **ascorbic acid**. Fresh fruit juices are often rich in Vitamin C, but orangeade, lemonade and other fizzy drinks do not usually contain any Vitamin C.

Fruit drinks add variety and colour to a meal and are refreshing. They are not essential in the diet, but water is. They add colour and flavour to water, but the water is still the most important part. Fizzy drinks usually contain too much sugar; it is wiser to drink water or a freshly made fruit drink containing Vitamin C.

orange juice — pure juice of fruit

fruit squash — 25% fruit juice

fruit drink — 10% fruit

fruit crush — 5% fruit juice

orangeade / fizzy drinks — no real fruit necessary

Things to do

1. Look at these two labels. What is the main ingredient in each? Which product contains most sugar?

   Ingredients: sugar, carbon dioxide, orange flavour, fruit acid, preservative, emulsifier, saccharin

   Ingredients: orange, pineapple, sucrose, fruit acid, sodium citrate, colour, saccharin

2. Design a poster to advertise your own home-made fruit drinks.
3. Draw and label four citrus fruits.
   Fill in the missing words:
   a) All citrus fruits contain plenty of ——————— —.
   b) Lemon tastes ————.
   c) Grapefruit tastes b—————.
   d) Orange tastes s—————.
   e) Lime tastes s———.
4. Rearrange the list, starting with the drink which contains the most Vitamin C.
   a) orange crush b) orange juice c) orange drink d) orangeade.
5. Which drink is made by infusing the leaves of a plant in boiling water?
6. Compare the cost of 1 litre of home-made lemon drink with 1 litre of bought orange juice.

New word
ascorbic acid – chemical name for Vitamin C

Fresh lemon or orange drink

Ingredients
(for 2–3 people)
1 lemon or orange
2 tblsp granulated sugar
600 ml water

Utensils
vegetable peeler, saucepan, lid, wooden spoon, lemon squeezer, tablespoon, measuring jug, glass or serving jug

Method

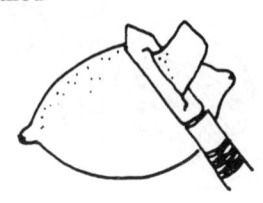

1 Wash and dry the fruit. Remove the **zest** only using a peeler.

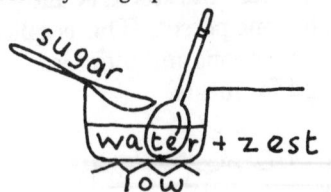

2 Place sugar, water and zest in saucepan. Stir over a low heat to dissolve sugar, until liquid becomes clear.

3 Leave the syrup to cool. **Infuse** the zest of the fruit in the covered pan to draw out the flavour.

4 Squeeze the juice from the fruit into a jug and poor on the cool syrup. It must be cool so that it does not destroy the Vitamin C in the fruit juice.

5 Serve chilled. Decorate the edge of the glass with a slice of the fruit.

Citron pressé

Ingredients
(for 1 person)
1 lemon
300 ml water
2 tblsp sugar

Utensils
lemon squeezer, knife, spoon, glass, jug

Method

1 Wash the lemon, cut in half.

2 Squeeze the juice into a glass. Add the sugar.

3 Stir with a long-handled spoon to dissolve the sugar. Serve the water in a separate jug, adding as much as is needed to taste.

Plenty of Vitamin C is kept in these two fruit drinks.

> **New words**
> **infuse** – to soak in a liquid in a covered container to draw out flavour
> **zest** – thin outer skin of citrus fruits, it contains the flavour of the fruit so is useful in cooking

Timing and table laying

When making a meal, several foods may be prepared and cooked together. These foods may all need different cooking times, e.g. sausages take longer to cook than tomatoes, meat loaf takes much longer to cook than frozen peas. All the parts of a meal should be ready to serve at the same time; it would be unpleasant if the tomatoes were burnt while the sausages were undercooked. The easiest way to make sure the food is cooked correctly and served hot at the right time, is to work out how long each food takes to prepare and cook e.g.

Food	Prep. time	Cooking time
sausages	0	20 mins.
bacon	3 mins.	10 mins.
tomatoes	2–4 mins.	5 mins.

Start by cooking the food that takes the longest, and add the other items at the right time.
11.40 begin cooking sausages (20 mins.)
11.50 add bacon (10 mins.)
11.55 add tomatoes (5 mins.)
12.00 serve.

Now the meal is ready to serve, how is it to be eaten? Think about the **cutlery** needed to eat the meal and the **crockery** to serve it on. For a meal of bacon, sausage, tomato, toast and tea, these will be needed: a plate to eat from, a knife and fork to eat with, a side plate for the toast and a small knife to spread it with butter, a napkin, salt and pepper, a cup and saucer or tray laid for tea.

The diagram shows where the crockery and cutlery are placed for one person. This is called a place setting. Other spoons and different items may be needed for meals.

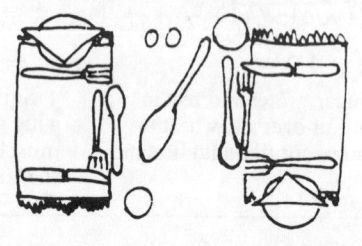

A table cloth or place mats may add to the attractive appearance of the table as well as protecting the table top.

Things to do

1 The diagram shows a place setting for the meal given opposite. There are many mistakes. Correct them and draw the correct place setting, adding any missing items.

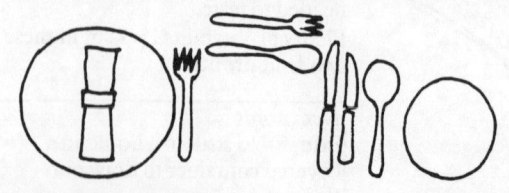

2 Draw a tray laid for tea.
3 Make a list of the crockery and cutlery needed to lay the table for the following meals. Draw a place setting for one of them.

a) stuffed egg, salad, bread and butter, peach melba, coffee
b) soup, Welsh rarebit, apple
c) baked stuffed fish, mashed potato, tomato sauce, lemonade, fruit salad

4 Prepare time plans for the following meals:
a) Baked stuffed fish, boiled potato, frozen peas, grilled tomato, lemonade, pear jelly
b) Vegetable soup, scrambled egg, toast
c) Sausage hot pot, baked potato, upside down cake

New words
cutlery – knives, forks and spoons with which food is eaten
crockery – plates, bowls and china off which food is eaten

Preparing a meal — breakfast

grapefruit, grilled bacon, sausage, tomato, toast, tea

Ingredients
(for 2 people)
1 grapefruit, 2 tblsp caster sugar
4 sausages
4 rashers bacon
2 tomatoes
salt, pepper, oil
2 slices bread
2 tsp tea, milk, sugar

Utensils
scissors, grapefruit knife, brush, tablespoon, tongs, fork, vegetable knife, serving bowls and plates

Method
Collect ingredients, utensils, cutlery and crockery.

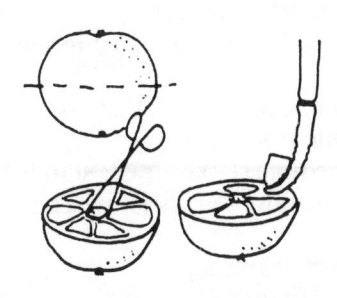

1 Prepare the grapefruit: cut in half, cut out centre core with scissors. Use a serrated or grapefruit knife to loosen each segment so that it comes away from the skin. Place in serving dishes, sprinkle with sugar, chill.

2 Lay the table, put serving dishes to warm.

3 Line the grill pan with aluminium foil to make cleaning easier. Adjust the grid to the correct height. Make the toast, keep warm.

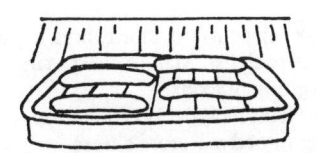

4 Arrange sausages on grid, cook for 10 minutes on medium heat.

5 After 10 minutes turn sausages, using tongs or 2 spoons. Add the bacon. Cook for 5 minutes.

6 Turn the bacon after 5 minutes. Check sausages are browning evenly.

7 Cut tomatoes in half. Make an X cut in the centre. Brush with oil, season.

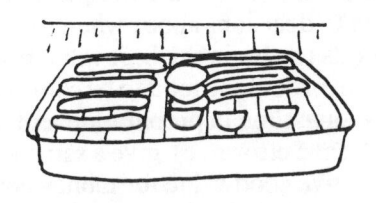

8 Place tomatoes under the grill for last 5 minutes.

9 Boil the kettle, make the tea. Serve the meal on warmed plates.

61

Convenience foods

The term convenience food is used to describe foods that need little or no preparation, e.g. peas shelled for frozen peas, bread sliced for sliced bread, meat cooked in tinned meat pies.

Complete meals can be bought frozen, in tins or dried and sold in packets.

Many convenience foods are used so often in the daily diet that they are taken for granted – frozen vegetables, custard powder, stock cubes. However, it is often useful to know what to do if the convenience food is not available, e.g. how to prepare vegetables, make custard and prepare stock. Some convenience foods are expensive to use, they may be unnecessary and not really very convenient at all.

The consumer must pay for the preparation to make the food more convenient, so it is important to be able to choose sensibly and make decisions about when to use convenience foods to obtain the best value from them.

In order for a food to be convenient it should:
a) save time in preparation and cooking
b) save effort   c) give a satisfactory result
d) give good value for money compared with home-made food.

Prepared meals and snacks from 'take-away' food shops should also be considered with home-made and other convenience foods.

Food additives

Most prepared foods contain added chemicals, or food additives. The home cook does not need to use chemical additives. But they are used in tinned, frozen and convenience foods to keep them fresh, to add or improve colour and flavour, and to give the special qualities of the foods.

Food labelling regulations state that all ingredients must be listed on the label in order of weight, including food additives. Read the labels on tins and packets of food, some additives often used are:

a) monosodium glutamate (m.s.g.) – in savoury foods to improve flavour
b) hydrolised vegetable protein – to improve meaty flavour
c) textured vegetable protein (T.V.P.) – artificial meat made from soya
d) food starch or edible starch – pure starch
e) saccharin and sorbitol – artificial sweeteners
f) sucrose – sugar
g) caramel – brown colouring
h) ascorbic acid – Vitamin C

Things to do

1 Compare home-made hamburgers with frozen, and two types of take-away hamburger.
Fill in the charts with the results:

	Time to prepare	Time to cook
A home-made		
B frozen		
C take-away		
D take-away		

List of ingredients	Appearance/flavour	Cost
A		
B		
C		
D		

Which hamburger contains most meat?
Is there any difference in size and weight before and after cooking?
Which hamburger do you prefer?

2 Many convenience foods can be used attractively to make up dishes, and are useful to keep in store. Choose foods from this list that would be useful to have in store. Suggest a use for each, e.g. tinned mince to make shepherds pie.
   packet cake mix, tinned mince, tinned stewing steak, tinned fruit pie filling, dried curry meal, instant mashed potato, frozen pastry, frozen vegetables, ice-cream, frozen meat pies

3 Find out how to make egg custard, stock, bread, gravy.

4 Compare the foods in each list. Make a chart as in question 1 for each group of foods.
   a) fruit fool, bought fruit fool, packet fruit whip
   b) macaroni cheese, tinned macaroni cheese
   c) muesli, different makes of packet product
   d) mashed potato, instant mash

5 Collect labels and packets of convenience foods. Look at the list of ingredients and find out the additives.

Hamburgers

Plan of work (hamburger, salad, milk shake)
1. Collect ingredients and utensils
2. Lay the table
3. Prepare milk shake. Chill.
4. Prepare hamburger mixture.
5. Prepare salad. Cover and chill.
6. Make dressing.
7. Fry hamburgers. Dress salad.
8. Serve.

Ingredients
(*for 4 hamburgers*)
200 g mince or minced chuck steak
1 small onion
salt, pepper
cooking oil
flour
4 soft bread rolls

Utensils
frying pan, basin, grater, plate, wooden spoon, tongs, fish slice, flour sifter, serving plates

Method

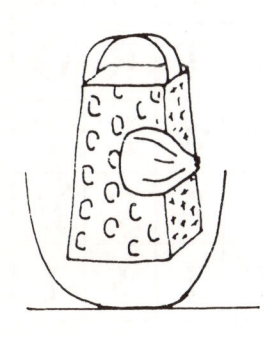

1 Grate or finely chop the onion.

2 Mix onion, meat and seasoning in a basin.

3 Divide into 4. Use floured hands on a lightly floured table to shape into 4 cakes about 1 cm thick.

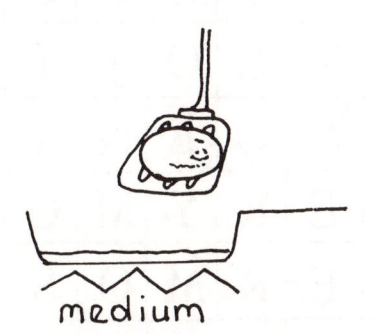

4 Cover the base of frying pan with oil. Heat gently until a cube of bread browns in 1 minute. Lower the hamburgers into the hot fat. Cook for 5 minutes on each side, turning carefully with tongs or 2 spoons.

5 Drain on kitchen paper. Split the rolls and sandwich each hamburger in a roll.

Serve with a mixed salad and milk shake.

Magic word square

Here is a magic word square. In it are hidden all the new words you have learnt in this book. Find as many as you can and write them in your exercise book, putting each one on a new line. When you have finished, write the meaning beside each word. (The 'New Word' boxes in the book will help you, if you find this difficult.)

Remember – the words in the square might read across, down, up, backwards or diagonally in any direction!

M	L	D	W	S	R	A	D	I	A	T	I	O	N	Y
M	I	O	G	E	I	N	F	U	S	E	C	T	A	A
N	J	C	O	A	G	U	L	A	T	I	O	N	B	P
O	U	U	R	S	R	Q	E	U	G	F	N	E	B	R
I	E	T	P	O	N	N	O	Z	O	W	V	T	R	O
T	S	L	R	N	N	H	I	O	E	T	E	W	E	T
C	U	E	I	I	C	U	R	S	N	S	C	E	V	E
U	F	R	O	N	T	P	T	D	H	U	T	I	I	I
D	N	Y	A	G	N	I	A	R	G	J	I	G	A	N
N	I	L	Y	E	I	E	O	J	I	S	O	H	T	P
O	B	W	V	X	N	C	E	N	I	E	N	T	I	R
C	R	O	C	K	E	R	Y	M	C	D	N	G	O	O
F	G	E	T	H	E	R	M	O	S	T	A	T	N	U
H	Y	G	I	E	N	E	E	R	U	P	H	E	S	X
Y	A	S	C	O	R	B	I	C	A	C	I	D	A	D

64